Transferring Food Production
Technology to Developing Nations

Westview Replica Editions

The concept of Westview Replica Editions is a response to the continuing crisis in academic and informational publishing. Library budgets for books have been severely curtailed. Ever larger portions of general library budgets are being diverted from the purchase of books and used for data banks, computers, micromedia, and other methods of information retrieval. Inter-library loan structures further reduce the edition sizes required to satisfy the needs of the scholarly community. Economic pressures on the university presses and the few private scholarly publishing companies have severely limited the capacity of the industry to properly serve the academic and research communities. As a result, many manuscripts dealing with important subjects, often representing the highest level of scholarship, are no longer economically viable publishing projects--or, if accepted for publication, are typically subject to lead times ranging from one to three years.

Westview Replica Editions are our practical solution to the problem. We accept a manuscript in camera-ready form, typed according to our specifications, and move it immediately into the production process. As always, the selection criteria include the importance of the subject, the work's contribution to scholarship, and its insight, originality of thought, and excellence of exposition. The responsibility for editing and proofreading lies with the author or sponsoring institution. We prepare chapter headings and display pages, file for copyright, and obtain Library of Congress Cataloging in Publication Data. A detailed manual contains simple instructions for preparing the final typescript, and our editorial staff is always available to answer questions.

The end result is a book printed on acid-free paper and bound in sturdy library-quality soft covers. We manufacture these books ourselves using equipment that does not require a lengthy make-ready process and that allows us to publish first editions of 300 to 600 copies and to reprint even smaller quantities as needed. Thus, we can produce Replica Editions quickly and can keep even very specialized books in print as long as there is a demand for them.

About the Book and Editors

*Transferring Food Production Technology
to Developing Nations:
Economic and Social Dimensions*
edited by Joseph J. Molnar and Howard A. Clonts

This book explores the social, economic, and policy problems associated with introducing new agriculture and aquaculture technology to developing nations as a means for expanding food supplies and increasing well-being. The contributors examine three general facets of planning for technology transfer and consider methodologies that enable effective integration of social and economic factors. The first section of the book covers problems of planning at the national and regional level, emphasizing methods and models for macro planning under conditions when resources are limited. Two subsequent sections, focusing on planning at the local level and on constraints on the technology transfer process, cover a broad range of topics, among them production and marketing decisions by small farmers, conflicting objectives of planners and producers, limitations of resource allocation within the production unit, and strategies for training extension workers, researchers, and project planners.

Joseph J. Molnar is associate professor of rural sociology at Auburn University. Associated with the International Center for Aquaculture, he teaches courses in extension methods, research methods, and the sociology of natural resources and the environment. *Howard A. Clonts*, professor of resource economics at Auburn University, serves as departmental liaison to the International Center for Aquaculture. His research and teaching areas include aquacultural economics, project planning and evaluation, water resource management, resource economics, and resource policy.

Transferring Food Production Technology to Developing Nations

Economic and Social Dimensions

edited by Joseph J. Molnar
and Howard A. Clonts

Westview Press / Boulder, Colorado

A Westview Replica Edition

Published in 1983 in the United States of America by
     Westview Press, Inc.
     5500 Central Avenue
     Boulder, Colorado 80301
     Frederick A. Praeger, President and Publisher

Library of Congress Catalog Card Number: 83-50067
ISBN 0-86531-957-X

Printed and bound in the United States of America.

10 9 8 7 6 5 4 3 2 1

Contents

PART I

AGRICULTURAL DEVELOPMENT AND TECHNOLOGY TRANSFER

PART II

MACRO-LEVEL PLANNING IN DEVELOPMENT

PART III

MICRO-LEVEL PLANNING IN DEVELOPMENT

PART IV

SOCIAL AND INSTITUTIONAL CONSIDERATIONS

Tables and Figures

Acknowledgments

While we bear primary responsibility for assembling the authors who contributed to this volume and for its editing, the results are truly the effort of a much larger group of individuals. An initial grant from the Dean of General Extension, Dr. Gene Bramlett, enabled the project to get underway. Generous support from the Southeast Consortium for International Development allowed a number of scholars, who would have been unable to attend otherwise, to travel and participate in the conference.

Several far-sighted administrators fostered and encouraged the endeavor. Dr. Stanley P. Wilson, Vice President for Agriculture, Home Economics, and Veterinary Medicine and Dr. Gale A. Buchanan, Dean and Director, Alabama Agricultural Experiment Station, were especially supportive of efforts to increase the on-campus supply of social science expertise for the International Center for Aquaculture (ICA). Dr. Wayne Shell of the Department of Fisheries and Allied Aquacultures and Dr. Joseph H. Yeager of the Department of Agricultural Economics and Rural Sociology were particularly effective in managing the continuously complex administrative matters associated with interdepartmental involvement in the ICA. As Associate Director of the ICA, Dr. Donovan Moss was encouraging and helpful throughout the process.

We are particularly indebted to John B. Flynn, who took on important responsibilities in the early planning and organizational phase of the conference. Now having resumed a foreign service career, Auburn's loss is USAID's gain.

ICA staff members made significant contributions to the effort that resulted in this volume. We thank professors Rudy Schmittou, John Grover, Brian Duncan, and David Rouse, as well as Ron Phelps and David Hughes, for their active support and enthusiasm. We also thank John Dunkelberger and others who commented on our manuscript.

Leisha McCarty, Carrie Rybarczyk, and Rhonda Crews made major typing contributions to this volume, often under trying circumstances. Dan Tamblyn became our resident expert on the University of Chicago Manual of Style and did most of the indexing. Without their cheerful and competent assistance, the task of bringing this book to completion would have been a much longer and more difficult task.

Joseph J. Molnar
Howard A. Clonts

1

Technology as a Source of Economic and Social Advancement in Developing Countries

Joseph J. Molnar
Howard A. Clonts

Perhaps the most pervasive feature of efforts by industrialized nations to assist the developing world has been the introduction and diffusion of new and better ways of growing plants and animals.[1] Because exponential population growth rates loom in most developing nations, government officials and planners look to a self-sufficient agricultural sector as a first step toward a better future.[2]

World population growth dictates in large measure the increases needed in food production. Since the beginnings of agriculture, world population has increased 256-fold (eight doublings), and is now nearing 4.5 billion people.

World population reached 1 billion in the period from roughly 12,000 B.C. until 1850. It took only 80 years to reach 2 billion, and only 45 years to reach 4 billion. To simply maintain today's already inadequate per capita food consumption levels is an awesome challenge by itself. The world food supply must be doubled again by the first decades of the 21st century. The magnitude of the food production tasks that lie ahead requires that we find ways to accelerate the process of generating and diffusing food production technology throughout the world.[3]

While some point to inherent limitations on the land, in most regions of the world farms are potentially far more productive than present practices permit. Existing technology can greatly increase production. The promise of agricultural technology challenges political leaders and economic institutions to extend the benefits of expanded productivity to the developing nations.[4] Furthermore, failure to extend modern technology may damage future productivity through soil erosion, desertification, and other consequences of wasteful grazing and cropping practices. As population growth and industrialization are generating pressures to convert farmland to nonfarm uses, the net effect may be a loss of momentum in the growth of world food production.[5]

The purpose of this chapter is to provide an overview of the technology transfer process and its role in international development. Technology transfer is not a new concept, but it is widely misunderstood and frequently misapplied. One objective is to examine technology transfer as it occurs on a societal level through intentional government investments over and above natural diffusion processes. A second objective is to review the determinants of success in technology transfer on the farm and village level, particularly the development of domestic capacity to receive technology and adapt it to local needs and conditions. Finally, social and political barriers are considered in light of their significant potential to mitigate or magnify the success of otherwise well-conceived technology transfer efforts.

TECHNOLOGY TRANSFER AND DEVELOPMENT

The pace of world population growth demands an active rather than a passive response to the need for accelerated agricultural productivity in developing countries. Technology is the organized capability for some purposeful activity.[6] The transfer of technology is a purposive aspect of the broader process of cultural diffusion and is not limited to one sector of the economy or another. That is, new ideas, practices, and objects are constantly being exchanged in an undirected yet active way in the ongoing intercourse of communication, travel, trade, and migration among and between societies.[7] In technology transfer, however, ways of doing things that are somehow new, better, or different are purposively transmitted from one setting to another with the intent of affecting that setting in a number of direct and indirect ways.[8]

Technology is not transferred in a single identifiable event. Rather it is a process, an evolution, a concomitant aspect of economic growth.[9] Innovations are introduced into a nation because some benefit is expected by the government, by an entrepreneurial individual or corporation, or by an outside agency seeking to promote an expanded supply of cheaper food.

Agricultural innovations generally involve a sequence or configuration of objects, skills, and facilities. The transfer often can be accomplished through the market place, given the intrinsic advantages, attractiveness, and availability of the innovation. In many developing countries, however, the technology may be neither directly available nor immediately applicable to the targeted setting or user group. Some technologies may never achieve widespread utilization because traditional peasant farming systems frequently lack the education, resources, experience, and incentives necessary to alter customary approaches and undertake new ways of producing food.

When market processes are unable to deliver change, or the pace of such change is unacceptably slow, governments attempt to foster the transfer of technology through an organized set of efforts termed development projects.[10] The mid-1960's marked an increased effort to extend improved agricultural technology to the developing world.[11] The establishment of 13 international agricultural research centers over the past two

2

decades has expanded agricultural research on major food crops and farming systems. Although the most impressive achievements to date have been in wheat and rice, efforts continue with other crops at places such as the International Potato Center in Peru. The introduction of high yielding wheat and rice varieties, in conjunction with improved agronomic practices, has allowed India to attain near self-sufficiency in food production in recent years.[12]

MACRO-LEVEL PLANNING PROCESSES

A development project is some well-defined activity, usually an investment, but not necessarily so, that the government of a developing country considers as contributing to its growth or the general welfare of the people. Development projects must be technically viable so that they fit the intended biological and physical setting, produce an output that is in demand, and meet the economic, financial, and other requirements of the participants.[13] Such projects are generally the mechanism for the purposive transfer of a new technology, but a lack of well-conceived projects historically has been a primary weak point in the development planning process.

Macro-level planning involves a broad perspective of the long-term implications of technology transfer. Strategies of comprehensive national or regional development encompass the concept of Pareto optimality for which the goal of a social welfare maxima is implicit. More frequently, welfare is measured in terms of quantifiable quality-of-life factors which are linked to political stability and enhanced individual freedom. Thus successes in macro planning frequently are measured in terms of actual improvements in the protein content of local diets, or the amount of foreign exchange generated in the agricultural sector.[14]

Technology transfer efforts are increasingly evaluated by a standard of success broader than increases in yield. For example, the seed-fertilizer technology developed for key cereal grains bypassed a significant proportion of agroclimatic environments and poorer farmers. Technology transfer also is held accountable to a variety of non-production goals including increases in overall rural employment, more equitable income distributions, and the development of local institutional capacity.[15] Thus the planning of development projects requires an assessment of a number of potential institutional, economic, and agronomic impacts of a new product or technology.

As Simpson points out in this volume there is considerable need to clarify visions and goals in development. The strategies developing nations use to generate visions and goals are often confused. This is particularly true in the case of efforts to promote food production and growth in the rural sector. Many developing nations have tended to emphasize economic growth without specifically considering the manner in which the

3

benefits of growth may be distributed. The expectation is that expansion in any economic sector will generate welfare gains which in turn will spread throughout the society. However, there is an increasing recognition among developing country governments that the first step toward a better future is a productive and efficient agricultural sector.[16]

Spurring Growth in Agriculture

Mosher argues that agricultural policy makers can undertake three simultaneous actions to spur agricultural growth, each of which directly involves the transfer of technology.[17] Improving the efficiency of regular agricultural agencies calls for a tune-up of the machinery by which technology is identified, adapted, and disseminated. A number of practical administrative concerns relate to the location of agricultural support services, the selection and training of extension personnel, and other operational matters that directly influence the sustained implementation of improved practices at the farm level.[18]

Commodity production programs attempt to increase the production of specific commodities by fostering the implementation of new technologies wherever they are effective and profitable. Mosher emphasizes the great importance of tailoring efforts aimed at accelerating agricultural growth to situations in different parts of the country.[19] Others have made the point in terms of the appropriateness of technology,[20] but both concepts center on the fit of the new practices, varieties, breeds, or farming system to local conditions, capabilities, and capacities. That is, transfer cannot be accomplished if the technology ultimately is not reconciled with the setting in which it is to be applied.

Mosher's third action to promote agricultural growth then is to focus research and extension activities on farming district projects that try new methods and technologies in limited geographic areas.[21] Such projects ensure that new technologies have the appropriate infrastructure of farm credit, production inputs, marketing facilities, and extension assistance while focusing on overall crop and livestock production in an area.

Production technology is adjusted to local conditions through on-farm testing, technical assistance to producers, adaptive research, and an interactive relationship between extension and research. A more recent approach focuses on the transfer of technology to a farming system, taking a holistic perspective on the interrelated nature of resources, labor, and markets. Farming system projects are interdisciplinary efforts intended to introduce and adapt technology in a way that interacts and responds to the agronomic conditions, family patterns, and economic needs of the indigenous agricultural producer.[22] Sustained research efforts are necessary to resolve the problems that always arise as such a new program progresses.

4

New technology should not be regarded as a miracle "fix" for the manifold problems of developing better, more productive arrangements for growing crops and animals. Technology alone cannot overcome the problems of overpopulation, resource deterioration and world economic imbalance. As Brady notes, however, there is some room for cautious optimism, as some countries have improved their productive capacities to the point of self-sufficiency and others are making remarkable strides. Technology and an international network of research centers have made fundamental contributions to the progress that has been made.

Policies and Institutions

A governing influence on the efficacy of new technology, however, is the policies and support mechanisms that allow the new methods and varieties to be implemented and successful. Land tenure systems, artificially administered prices, and the availability of competent and motivated technical assistance all affect the viability of an otherwise useful and productive technology.

Government policies influence agricultural development in a number of fundamental ways. The array of values or direction taken by a country's leadership results in a vision or intended kind of progress for the agricultural sector. Simpson argues that planning proceeds from a prior set of values and assumptions about development that translate into intermediate goals or objectives. The utility or realism of these objectives is a function of 20- or 30-year projections of economic and demographic conditions that establish a context for implementing new technologies. The complexity of new technology and the uncertainties of future conditions ensure that technology transfer is generally guaranteed to be a somewhat idiosyncratic process geared to the characteristics, needs, and timing of national efforts as well as the efficacy of the government's agricultural apparatus.

Institutions play key roles as conduits as well as barriers to the application of new technology. Storer reviews the case of ocean fisheries and the unique nature of fish as a food commodity to draw conclusions about fisheries development and management. Government policies and national institutions comprise a set of macro-level parameters that constrain or channel efforts to transfer technology and may in themselves be changed by the conditions resulting from the implementation of new technology.

Although fisheries planning is primarily a national concern extending to the 200 mile limit, the fish themselves are not bound by the boundaries. Protection of the scarce resource requires international collaboration. A growing technological capability to mine the world's fishery is accompanied by a growing need to protect the collective resource. In this case, new technology places new demands on unclaimed ocean space, raising significant questions about biological capacity to sustain yield and institutional responses to protect the resource.[23]

MICRO-LEVEL PLANNING PROCESSES

Central governments face complex problems in choosing and implementing large or small-scale projects that make sense to the peasant farmer and life in the village, and contribute as well to overall national objectives. The identification, design, and analysis of projects call into play the skills of multiple disciplines to assess the feasibility of an investment. Economic and financial analysis must proceed on the basis that activities provide adequate incentive to individual farmers as well as a return to the investors or government agency sponsoring the project.

Micro concepts, as already implied, require a more refined and limited application of macro-level tools. The major objective of technology transfer at the micro level is to maximize the likelihood that new technology will be absorbed by individual producers and consumers. Inadequate resources, credit constraints, lack of markets and market access, as well as limited educational preparation to absorb technical information all contribute to difficulties in initiating lasting change.[24] A major source of failure in the technology transfer process has been persistent attempts to install overly sophisticated production systems or simply to transfer scaled-down technologies that were not in tune with the needs, incentive system, or capacity of the recipient country.[25]

Small-scale projects must be technically sound but they also must meet stringent economic criteria which have linkages to central government plans. Failure to integrate individual projects into overall macro-economic social goals has been a reason for failure in an embarrassing number of cases.[26] Few countries have designed an overall plan for development and, even if partial planning for particular areas or regions may be more realistic in some cases, there is a clear need for linkages among sector programs and projects.[27]

Projects may transfer technology by contributing otherwise unavailable infrastructure to a local area, such as an aquacultural pond or an irrigation system. Projects also can provide a means for introducing improved methods of design, construction, or operation of facilities, now or existing, into the experience of the indigenous population.[28] Some projects, particularly those employing a farming systems approach, actually generate improved technology in the context of local conditions and constraints.

Technical Capacity

One problem related to the fit or appropriateness of a technology to a developing nation involves the sophistication

6

or complexity of new equipment or facilities. Problems associated with initial success of new farm machinery and the later failures due to lack of tractor repair are commonly known. Yet, many times larger production or processing facilities involving complicated technologies are necessary for more efficient internal distribution or international marketing of a product.

Automatic control devices often allow a developing country to easily start production with an imported facility provided it has trained the necessary operators. The trouble begins when the imported facility suddenly breaks down. Maintenance and repairs are always necessary, but the replacement of any one part requires a whole modern industrial base that can only be provided by developed countries. Thus a nation that fails to supply a domestic scientific and engineering capacity increases its dependence on both foreign technology and foreign exchange.[29]

The Hayami and Ruttan model of technology transfer points to three phases of activity reflecting the evolution of a nation's capacity to receive and alter technology for its own purposes.[30] Material transfer involves the adoption and diffusion of things such as seeds, plants, machinery, pesticides, and fertilizer.[31]. Design transfer activities are characterized by the inculcation of knowledge about new products and processes into the nation's experiment stations and is reflected in adaptive research and systematic testing on farmers' fields. The third stage, capacity transfer, reflects the process of institution-building in a developing country. The nation itself then possesses the trained personnel, scientific laboratories, and experience to generate new varieties, breeds, and machinery that will provide the base for productive self-sufficiency and eventual contributions at the frontiers of science.

The long-term objective of capacity transfer is what Dahlman and Westphal term "technological mastery."[32] Mastery of technology comes from effort, not transfer. The ability to make effective use of technology can only be acquired through indigenous effort. Accumulated experience leads to an enhanced institutional capability to adapt technologies better suited to local circumstances.

Technology mastery is directly linked to institutional development. Technology transfer extends to the institutional sector as well as to the farmer, for infrastructure is required to support and sustain the adaptation, introduction, and continued use of new technology. Host country institutions play an important role in filtering western technology, making choices about what will fit local conditions and what must be reworked to conform to labor-surplus, capital-short small farm systems. Axinn underscores the importance of a sense of partnership between those extending assistance and host-country

7

institutions, not only to ensure the success of technology transfer efforts, but to build capacity for the long-term.

Stevens points out that most agricultural areas in developing nations are still in the beginning phases of a century-long agricultural and rural transformation. Thus the potential impact of a new technology should be assessed in light of its short-term effects on farms and villages as well as the long-term dynamics of national development. Furthermore, the highly variable settings that are the developing nations makes the technology transfer process very much an endeavor shaped to the culture, climate, and needs of each country.

In many developing nations, low-cost labor and high-cost capital make capital-intensive agricultural technologies unprofitable for most farmers. According to the induced innovation perspective, much western technology has been developed in an environment of high-cost labor, thus "inducing" innovations by researchers serving a clientele facing this problem and ensuring ready adoption by the farmers with access to capital.[33] That is, farmers will seek to substitute their highest-cost input or use it more efficiently and the institutions serving this clientele will be led to produce innovations that facilitate the process.

In developing countries, labor is plentiful, capital is scarce. Technologies developed in a setting where just the opposite conditions prevail may not be appropriate for the peasant farmer. As Stevens indicates, the greater challenge in technology transfer is to articulate the more profitable technologies appropriate to the farming systems of low labor cost nations.

Production and Marketing Decisions

Roe explores T. W. Schultz's hypothesis that few significant inefficiencies exist in the allocation of the factors of production in traditional agriculture. Although new technologies may dramatically increase productivity, farmers tend to be risk averse and many times are unwilling to exchange predictable subsistence for uncertain abundance. How traditional agriculture can be transformed from a situation where farmers supply the bulk of their own inputs to one where they employ manufactured products is treated as a function of experience, cognitive ability, and access to information. Based on studies of Tunisian and Thai farmers, he points to the farmer's understanding of production characteristics of new varieties as determinants of input choices and the productivity realized from these choices.

Farmers make fairly wise choices based on their own experience and traditional knowledge.[34] Thus a new technology must be accompanied by an infusion of educational assistance to substitute for lack of experience with the operating parameters of the innovation. To the extent that the technology differs from traditional practice, extension assistance is required; and to the extent to which it is competently supplied, efficiency gains will be realized by the farmers.

Marketing and production decisions are inextricably linked to risk and uncertainty as one spills over the other and vice versa. Street and Sullivan overview the market system in developing nations as it influences the production process. Government buyers, private middlemen, and cooperatives are three main avenues for assembly and sale of products from the farm level.

Governments influence markets through the provision of roads, assembly facilities, and shipping points. When prices are regulated to the farmer's disadvantage, new technology will not possess the appropriate incentives to sustain diffusion. Farmers may then have little reason to supply more products beyond what meets their basic needs. Similarly, new marketing technologies may require extensive educational efforts to assure risk averse farmers that a season's efforts will not disappear into the accounts of unprincipled middlemen or the disorganized ruins of a marketing cooperative.

Efforts to transfer production technology that fail to coordinate the disposal of increased yields will not reach the ultimate objectives of sustained adoption and improvements in income and well-being. The many new high-yielding varieties of rice, wheat and other crops now available around the world often allow dramatic increases in production. Yet the basic economic problems of marketing coordination that plagued development projects for years still persist.[35]

SOCIAL AND POLITICAL BARRIERS TO DEVELOPMENT

Although "agriculture comes first" in the process of development, the organization of the food production system is interactively related to the coherence of the economic system and state apparatus with which it must coexist. A nation's agriculture also is directly influenced by international markets and the general health of the world economy. Due to the emergence of a world system and increased interdependence among nations, Flinn and Buttel argue that the western experience in agricultural development may be essentially nonreplicable in many areas of the Third World. They identify four categories of peripheral countries: rapid growth, oil exporting, agricultural and mineral export, and a final category of overpopulated, resource poor, largely agricultural economies. Although not mutually exclusive or exhaustive, they offer the typology as one means of understanding the diversity of the developing world.

The political economy of technology transfer is a concern because most western agricultural technology has not been scale-neutral. That is, it tended to benefit large farmers more than small farmers, and larger farmers were more ready to adopt from the start.[36] In addition, when land holding is con-

9

centrated or tenure arrangements are exploitive, technology transfer will not be a productive substitute for structural change and land reform when improving the lot of the poorest of the poor is the project's objective.[37]

One response to the need for farm-level technology for peasant agriculture has been farming systems research. Farming systems approaches facilitate the diversification of agricultural knowledge and the ability of peasants to utilize new technology with minimal disruption to traditional farming practices. Farming systems research, however, is expensive, labor-intensive, and may not produce the overall productivity of experiment station approaches. Countries and donors are faced with hard choices about the strategies and impacts associated with different development approaches.

The physical aspects and operational procedures of a technology cannot be separated from the social norms and attitudes that surround its application. This assertion is particularly germane to the social organization of irrigation development. Probably, the single most important factor in agricultural productivity is water. In many countries irrigation facilities are a collective resource, the successful use of which requires collaboration among the farmers drawing from a common supply.

Communities often manage their own irrigation systems in Asia. Coward shows how increasing attention is being given to involving farmers in design of local systems, as well as in their operation and maintenance since remote bureaucracies cannot effectively accomplish these tasks. The case of irrigation technology illustrates the importance of a "fit" between technology and social organization at the village level. In many other instances the focus of the fit is at the individual farm and family level. Similarly, the operation of ponds as a community aquacultural enterprise may represent a parallel set of considerations.

CONCLUSION

Technology transfer is neither a simple nor direct process. The chapters in the volume address various aspects of the transfer issues--what happens at the policy level, at the farm level as well as how technology affects the social and cultural fabric of a nation.

The continuing challenge that confronts developed and developing countries alike lies in identifying, grasping, and implementing the tools of technology to expand food production in the face of rapidly growing populations. The task of knowing what technology is available, what is appropriate for a particular nation, region, village, and type of farmer, and what technology needs inventing or adapting is a very complex one indeed.

Technology transfer must overcome a variety of constraints to have a positive impact on food supplies. Three primary sets of obstacles limit the successful application of new ways of growing plants and animals: agronomic, economic, and sociocultural.

Agronomic limitations have not been fully overcome. Soils, climates, and terrain vary widely around the world. Although the scientific base for food production methodologies is fairly developed, our understanding of principles and process does not automatically generate recipes and complete answers on the farm level for every combination of circumstances. Furthermore, production methods generated under one set of economic and sociocultural conditions do not automatically extend to other settings.

Economic factors set parameters on what combination of investments will produce income for farmers, food for the populace, and development for the nation. Sociocultural constraints define the setting for technology investments, the reception new technology is likely to receive, and the amount of inertia it must overcome to be implemented. Transferred technologies often are foreign in multiple senses of the word. The origins of the technology are from another nation, as are the project personnel, and the idea or world view underlying the technology. Thus technology transfer cannot be accomplished solely through advanced planning and anticipation by those extending the technology to another nation.

The signal mark of success in technology transfer may lie in the development of domestic capacity to adapt, to shape, and ultimately to appropriate technology. A nation's role in the technology transfer process should not be one of a passive receiver but as an active partner in the transaction. Technology transfer should provide answers to immediate needs, but its most useful consequences are related to training, institution-building, and the overall enhancement of capacity to address indigenous production problems through an effective in-country research and extension system.

Technological mastery--an autonomous ability to identify, select, and generate technology--is one long-term vision that should be an integral part of efforts to transfer food production technology to developing nations. The papers in the volume represent one effort to further understanding of this process.

REFERENCES

[1]S. Wortman and R. W. Cummins, Jr., To Feed This World (Baltimore: Johns Hopkins Press, 1978); C. Hanrahan and J. Willet, "Technology and the World Food Problem: A U.S. View," Food Policy 5 (1976):413-419.

[2]T. W. Schultz, Transforming Traditional Agriculture, (New Haven, Connecticut: Yale University Press, 1964).

[3]Norman E. Borlaug, "Contributions of Conventional Plant Breeding to World Food Production," Science 219 (11 February, 1983).

[4]Eric Lerner, "World Lags in Applying Crop Technology," High Technology 2 (1982); Terry N. Barr, "The World Food Situation and Global Grain Prospects," Science 214 (4 December, 1981).

[5]Lester R. Brown, "World Population Growth, Soil Erosion, and Food Security," Science 214 (27 November, 1981).

[6]Robert A. Solo, Organizing Science for Technology Transfer in Economic Development (East Lansing, Michigan: Michigan State University Press, 1975).

[7]Everett M. Rogers, Diffusion of Innovations, 3rd ed. (Riverside, New Jersey: The Free Press, 1983); Laurence A. Brown, Innovation Diffusion (New York: Methuen, 1981); Elihu Katz, Martin L. Levin and Herbert Hamilton, "Traditions of Research on the Diffusion of Innovation," American Sociological Review 18 (1963):237-252.

[8]A. T. Mosher, Three Ways to Spur Agricultural Growth (New York: International Agricultural Development Service, 1981).

[9]Sherman Gee, Technology Transfer, Innovation, and International Competitiveness (New York: John Wiley & Sons, 1981); Subatra Ghatak, Technology Transfer to Developing Countries: The Case of the Fertilizer Industry (Greenwich, Connecticut: JAI Press.

[10]Warren C. Baum, "The Project Cycle," Finance and Development 7 (1982):2-9; The U.S. Agency for International Development, Design and Evaluation of AID-Assisted Projects (Washington, D.C.: Training and Development Division, Office of Personnel Management, 1980).

[11]Carl E. Pray, "The Green Revolution as a Case Study in Transfer of Technology," The Annals of the American Academy of Political and Social Science 458 (1981):68-80.

[12]Borlaug, "Contributions of Conventional Plant Breeding to World Food Production."

[13]W. W. Shaner, Project Planning for Developing Economies (New York: Praeger Publishers, 1979); T. M. Arndt, D. G. Dalrymple, and V. W. Ruttan, Resource Allocation and Productivity in National and International Agricultural Research (Minneapolis: University of Minnesota Press, 1977).

12

[14]William Paul McGreevey, "Measuring Development Performance," in Third World Poverty, ed. W. P. McGreevey (Lexington, Massachusetts: D. C. Heath, 1980); G. Edward Schuh and Robert L. Thompson, "Assessing Progress and the Commitment to Agriculture," in Third World Poverty, ed. W. P. McGreevey (Lexington, Massachusetts: D. C. Heath, 1980).

[15]Bruce Koppel, "The Changing Functions of Research Management: Technology Assessment and the Challenges to Contemporary Agricultural Research Organization," Agricultural Administration 6 (1979); Francois Hetman, Society and the Assessment of Technology (Paris: Organization for Economic Cooperation and Development, 1973).

[16]World Bank, Rural Development, Sector Policy Paper (Washington, D.C.: World Bank, February, 1975); Allen D. Jedlicka, Organization for Rural Development (New York: Praeger Publishers, 1977).

[17]Mosher, Three Ways to Spur Agricultural Growth.

[18]Daniel Benor and James Q. Harrison, Agricultural Extension: The Training and Visit System (Washington, D.C.: The World Bank, 1977); John Russell, "Adapting Extension Work," Finance and Development (1981).

[19]Mosher, Three Ways to Spur Agricultural Growth.

[20]E. F. Schumacher, Small is Beautiful (New York: Harper and Row, 1973); Robert E. Evenson, "Benefits and Obstacles to Appropriate Agricultural Technology," The Annals of the American Academy of Political and Social Science 458 (1981):54-67; Martin Brown and Mikoto Usui, Choice and Adaptation of Technology in Developing Countries (Paris: Development Centre of the Organization for Economic Cooperation and Development, 1974), pp. 63-20.

[21]Mosher, Three Ways to Spur Agricultural Growth.

[22]John D. Hyslop, ed., "Farming Systems Research Symposium," (Papers delivered at USAID Farming Systems Research Symposium, Washington, D.C., December 8 and 9, 1980); David W. Norman, The Farming Systems Approach: Relevancy for the Small Farmer, MSU Rural Development Paper No. 5 (East Lansing: MSU, Department of Agricultural Economics, 1980); Chris O. Andrew and Peter E. Hildebrand, Planning and Conducting Applied Agricultural Research (Boulder: Westview Press, 1982).

[23]One response has been to encourage horizontal integration--nonfishing employment or employment in underexploited fisheries to attract fishermen out of their existing jobs if

not also out of their community--may help scale down an over-
worked fishery to the level that could be maintained; Donald
K. Emmerson, Rethinking Artisanal Fisheries Development:
Western Concepts, Asian Experiences, Staff Working Paper No.
433 (Washington, D.C.: World Bank, 1980).

[24]David C. McClelland, "The Role of Achievement Orientation in the Transfer of Technology," in Factors in the Transfer
of Technology, eds. W. H. Gruber and D. G. Marquis (Cambridge,
Mass: MIT Press, 1966).

[25]Richard S. Eckaus, Appropriate Technologies for Developing Countries (Washington, D.C.: National Academy of Sciences,
1977).

[26]Shaner, Project Planning for Developing Economies.

[27]World Bank, Rural Development.

[28]World Bank, Rural Development.

[29]Toshio Shishido, "Japanese Industrial Development and
Policies for Science and Technology," Science 219 (21 January,
1983):259-267.

[30]Yujiro Hayami and Vernon W. Ruttan, Agricultural Development: An International Perspective (Baltimore: The Johns
Hopkins Press, 1971).

[31]Brown, Innovation Diffusion; Rogers, Adoption and Diffusion of Innovation.

[32]Carl J. Dahlman and Larry E. Westphal, "The Meaning of
Technological Mastery in Relation to Transfer of Technology,"
Annals of the American Academy of Political and Social Science
458(1981):12-25.

[33]Evenson, "Benefits and Obstacles to Appropriate Agricultural Technology."

[34]Peggy F. Barlett, Agriculture Choice and Change (New
Brunswick, New Jersey: Rutgers University Press, 1982); Peggy
F. Barlett, A Critical Survey of the Literature on Farmers'
Decision Making (Washington, D.C.: Agency for International
Development, May, 1978).

[35]Borlaug, "Contributions of Conventional Plant Breeding
to World Food Production."

[36]W. S. Saint and E. W. Coward, Jr., "Agriculture and
Behavioral Science: Emerging Orientations," Science 197 (19

14

August, 1977):733-737; R. E. Just Jr., A. Schmitz, D. Zilberman, "Technological Change in Agriculture," Science 206 (14 December, 1979):1277-1280.

[37]Alain de Janvry, The Agrarian Question and Reformism in Latin America (Baltimore: Johns Hopkins Press, 1981).

2
International Technology Transfer

Nyle C. Brady

There is a tendency these days to overlook the strategic importance of technology development and transfer in enhancing national economic growth and to assume that massive resource transfers are all that are required to help our less fortunate Third World neighbors solve their problems. All too often assessments of past economic development ignore the strategic importance of technology development and transfer as the foundation for much of the progress already achieved in agriculture and health in many of the developing countries.

I am pleased that the agency I serve--the U.S. Agency for International Development (USAID)--includes among its primary goals the promotion of technology and development and is currently placing greater emphasis on these processes than has been the case in the past. There is growing recognition that the development and utilization of improved technologies is essential for long-term progress in every low income nation with whom we deal.

CURRENT DEVELOPMENT PROCESS

As we meet today, the world economic situation for both developed and developing nations is less than optimistic. Growth rates in the developed nations have declined in real terms for the past 5 years. For 1982, their average growth rate is less than 1 percent. Growth rates in the developing countries, while somewhat higher than those of industrial countries, nevertheless have exhibited a similar downward trend. The low-income developing countries reported an average of 3.9 percent growth in 1981, but this is considerably lower than their growth of 5.9 percent in 1980 and with their longer-term growth of 4.5 percent between 1973 and 1980. These have been difficult times for developing countries.[1]

Population
Assessments of current economic development progress in most developing countries must begin with a consideration of population levels. The dramatic increases in population numbers stimulated by improved health and medical services follow-

ing World War II have had and are still having significant constraints on economic development of the low-income countries. Today's population of some 4.2 billion people is increasing by 70 to 80 million annually, and four of five of the new births occur in the developing world. These newborn babies must be fed, and future economic and social opportunities provided for them.

Even though percentage increases in population numbers have declined slightly in some developing countries in the past decade, the absolute number of births continues to rise and will likely do so for some time to come. This is because 40 percent of the people in the developing countries are 15 years of age and younger. This "population bulge" will move into child-bearing ages in the coming decades, giving steadily increasing population numbers. In practical terms, it means that for every two people in the globe today, at least an additional person will be added by the year 2,000--compounding the problems of finding food in a world of shrinking resources. Even if moderate success is achieved in population control and overall development, it is likely that the additional food that must be produced by the year 2020 will approach the total amount of food that is being produced today.

While the gloomy forecast of worldwide food deficits made in the 1960's and 1970's have not come to fruition, food supplies remain a critical problem for most low-income countries. Overall, they can look with pride on progress of the past 30 years during which their food production was about doubled, a remarkable feat. But per capita food production has edged up only slightly worldwide, and in some areas such as Africa south of the Sahara, it has actually declined. Per capita availability of beef peaked in 1976, grain in 1978, fish catch in 1970 and oil in 1973. The war on hunger is obviously far from being won.

When food requirements cannot be met by national production, the developing countries are compelled to import food with or without outside assistance. For example, since 1975, Mexico has been spending much of its valuable foreign exchange on importing corn, rice, beans, and wheat--the essential elements of the national diet. In 1980, due to a shortfall of the 1980 harvest the total agricultural trade balance in Mexico was a debit of $700 million[2]. Newspaper reports of the past few months call our attention to similar problems faced by India. Widespread droughts in 1982 are forcing India to import cereals just 1 year after its record production of wheat and rice.

In addition to population pressure, another stress on the international food situation are the increased demands due to rapid growth in per capita incomes in several middle income nations. Included are the major oil exporters such as Indonesia, Nigeria, and Venezuela, as well as other developing nations which have made rapid economic gains in the recent past such as Korea, Taiwan, Brazil, Malaysia, and Thailand. As peo-

17

ple become more affluent, they not only want to eat more food, but food of better quality such as meat and poultry. The extra grain and fodder needed to feed the animals must be produced.

Another factor has emerged to further complicate the world food demand pattern. The centrally planned economies, particularly the Soviet Union and China, have stepped up significantly their imports of cereal grains to meet the growing political pressure for more and better quality food. While some of this import demand is in response to low yields due to bad weather, analysts generally agree that the demand may well continue for some time into the future.

All of these factors are influencing current world food supplies and the demand for them. Critical analyses suggest that stresses on these supplies will continue at least for the next 2 or 3 decades. If the demands cannot be met, world food prices will remain high and will likely go higher, placing in jeopardy the truly poor nations and poor people in those nations unless indigenous food production can fill the gap.

Resource Deterioration

Problems in developing nations of deteriorating soil and water resources, such as desertification and erosion, seriously impede programs designed to improve food production and add to the cost of the food. For example, the centuries-old practice of "slash-and-burn" which involved cropping the land for 2-3 years and then allowing it to fallow for 8-15 years was reasonably stable. This system depended upon nature to cycle plant nutrients from deep in the soil through trees and underbrush to the surface where they are subsequently available for food crop production. While there is ecological merit to this time-consuming process, population pressures have forced a shortening of the time sequence for the rotations, leaving the land in fallow for only 3 to 5 years. This is inadequate time for the soil "regeneration," with consequent reductions in yield and increases in soil erosion potential.

Overgrazing of grasslands in low rainfall areas and over-cutting of trees for fuel wood are other practices which are threatening the natural resource base, especially in Africa. The Sahel drought of the seventies called the world's attention to the fragility of the food production system in that area. The problems are not only physical and biological but anthropological and social as well.

Guarded Optimism

Even though the present economic situation of the developing countries is unacceptable, and the demands by the growing world citizenry for food exceeds the capacity of many of the nations of the developing world, overall there is reason for guarded optimism. Progress, to a limited extent, has been achieved. Some countries have had remarkable success, and potentials and plans for further progress for increased food production are in evidence. Because of the adoption and application of improved technologies, practices, and policies, the world's farmers produced twice as much food in 1980 than they did in 1950. Moreover, the world is still capable of meeting substantial increases in global demand for food. The most serious danger is not a global food shortage, but an array of localized problems within Asia, Latin America and Africa that will require continuing attention.

TWO CRITERIA FOR SUCCESS

One cannot help but be impressed with the increased agricultural production, especially of cereal crops in the developing countries during the past 20 years. These increases were the inspiration in the late sixties for the term "Green Revolution" to describe the process of which they were a part. They were based on two major advances. First, new varieties and associated production technologies were made available to farmers, and in turn, were adopted by these farmers, Second, sources of chemical and financial inputs, and government policies were such as to make it profitable for the farmers to adopt these varieties and associated production technologies.

The question can be raised as to why this two-component development process had not surfaced earlier. Failure of governments, both developing and more developed, to focus attention on agriculture is one factor. Likewise, failure to recognize that western temperate-climate technology could not be transferred directly to the tropics with any degree of success is another one. A third is the failure to assess the significance of trained indigenous person power to a successful development process. A fourth is that at no time in the past had technologies been available which were sufficiently superior to the traditional ones as to truly attract the cultivators.

IMPROVED TECHNOLOGIES

When the modern wheat and rice varieties on which the green revolution was based first became available, they spread very rapidly to geographic areas for which they were suited. Small-scale low-income farmers who had been described as being too tied to their traditional agriculture suddenly became innovators. They abandoned their traditional varieties and practices, and adopted new ones fully as quickly as had the midwest farmers when hybrid corn became available in the 1930's and 1940's. They soon discovered that it was to their advantage to change. They could make more money by doing so.

The situation just described had not prevailed prior to the late 1960's. When I first went to the Philippines in 1953-55 as a young Cornell professor I was depressed by the prevailing, but fallacious concept that the development of

19

improved technology through research was not needed. Offici-
ally, the U.S. agency which supported us was not permitted to
get involved in research. The prevailing concept was that the
U.S. and other western nations already had the necessary technology on which development in the tropics could be based.
According to this notion, all we needed to do was to transfer
our technology to them. My 2 years in the Philippines in the
fifties convinced me that the development of technology suited
to the tropics along with person power development was the
first step to increased food production in these areas.

Let me give you another example. In the 1960's the Ford
Foundation encouraged the Indian Government to establish
"intensive agricultural districts," which were provided packages of technology. The assumption was made that existing
technology, along with plenty of fertilizer and financial
inputs, and training for farmers would make India's production
really take off. While some notable successes occurred, in
general the expected production increases did not take place.
It became obvious to all concerned that the technology being
sold to the farmer was not greatly superior to that which he/
she was already using. They began to understand that development was not a phenomenon utilizing merely the technology coming from outside the developing countries and imported for
their use. Rather, development would work only when the technology used was superior to that the farmer was already using
and was suited to the physical, biological and socio-economic
environments where the development was to occur.

Further, the leaders realized that while some of the technological components of the development process could be imported and used with little modification, most had to be created in the local environments where they were to be used, and
all components had to be tested and molded to fit these environments.

Technology--Mechanical versus Biological
One of the primary motivations for the technological revolution which took place in the 19th and 20th centuries in
Europe and the United States was the need to save labor. The
result was to replace human power and even animal power with
machines. The degree to which output per person per hour was
increased was used as a measure of success of the technology.
Labor was, and still is, expensive in the more developed countries.

In the developing countries labor is plentiful and cheap.
In contrast, machines are expensive both to purchase and to
operate. As a consequence, the kind of technology needed for a
2 hectare rice farmer in Indonesia is only remotely related to
that needed on an Iowa or California corporate farm. The Indonesian cultivator can be helped most by simple technologies
suited specifically to his environment. A 5-horsepower tiller
will take care of most of his mechanical needs. Varieties with

20

built-in resistance to insects and diseases which reduce his dependence on pesticides would be high on his "want" list. His social and economic requirements must be met, not those of the Iowa farmer.

Increased Emphasis on Research
One of the consequences of the Green Revolution was world-wide increased interest in and support for agricultural research for technology development. This interest was particularly evident in actions of donors, both national and international, and of developing country political leaders, research scientists and farmers. Problem-oriented science was recognized as an effective development tool. For the first time, an agricultural field scientist was given a Nobel prize. Scientists were honored by their governments for products coming from their research. Farmers began to have confidence in the products of the researchers and were willing to listen to extension personnel as the new technologies were spread. And, perhaps most importantly, additional financial resources were made available to support research by national and international donors and by developing country governments.

Among the international agricultural research efforts, probably the most successful is the network of international agricultural research centers sponsored by the Consultative Group on International Agricultural Research (CGIAR). Some 13 centers, each focusing on one or more specific development constraint(s), have been established. In addition to the research carried out at the headquarters of these institutes, collaborative networks have been established to assure coordination with national research systems. In turn, donors and developing country governments have given increased attention to the strengthening of national agricultural research organizations. Although the process has only gotten well underway, already improved varieties and associated technologies are products of these combined efforts.

The approach of the CGIAR network and that of other successful international science and technology efforts has seven common characteristics.

- Focus is placed on removing real constraints to development as opposed to research abstracted from practical needs.

- Funding has been through voluntary donations rather than by assessed contributions.

- Direct bilateral relationships have been encouraged between donors and recipient nation scientists and research institutions.

21

- Decisions have been based on development needs and technical merit.

- Existing mechanisms were used and strengthened for collaboration between developing countries and donors in the design, implementation, monitoring and evaluation of projects.

- A large bureaucracy with functions that overlap those of existing national and international entities was avoided.

- The needs of the lower-income developing countries received the highest priority.

To complement and enhance technology development by international centers and national research organizations, the United States has passed legislation which makes possible extensive use of American universities in both agricultural research and extension. In the long run, educators and scientists in the developing countries must take primary responsibilities for providing technologies suited to their national needs. But a serious lack of trained persons, of institutions, and of research and extension capabilities makes it necessary for assistance to be provided from the outside. Over the past three decades U.S. universities have been found to be well suited to provide much of that assistance.

Specifically, U.S. universities provide three kinds of help:

(1) Formal training and research experiences in America for thousands of developing country students.

(2) Technical assistance in agriculture to developing nations. This involves overseas participation by U.S. faculty members in collaborative projects of institution building, human resource development and specific research and extension products. USAID provides about $300 million annually for support of projects with which American universities are associated.

(3) Partnership research efforts between American scientists and counterparts overseas focused on problems which neither partner could solve alone, but which are amenable to solution through joint research.

The U.S. universities have the scientific expertise, sophisticated equipment and research methodology needed to tackle researchable problems which overseas farmers and, in turn, scientists face. The overseas partners can provide a testing ground for products or research and through cooperative efforts, can develop their own capabilities to do the more sophisticated research.

POLICIES AND SUPPORT MECHANISMS

Up to now I have stressed the singular importance of the development of truly superior technologies in enhancing the food-producing capacities of nations. But the facts are that the "environment" in which these technologies are to be used must be such as to assure benefits for the producer if he/she accepts them. If this condition does not pertain, the technology will not be adopted. It will be of value only to the researcher and his/her colleagues.

The "system" must enable the technology to work. Government policies and support mechanisms can foster farmer adoption and application of improved technologies. Farmers around the world--both developed and developing countries--will not adopt new technologies (regardless how good) under adverse economic conditions. For example, American farmers generally have adopted a new technology on a large scale only when the farmers are convinced it is to their economic advantage to adopt it. The same holds true for farmers of the Third World; they will not adopt improved technologies unless the price they receive for their harvests are set high enough to permit them to make a decent profit.

Additionally, farmers need access to credit, they need adequate roads to bring their produce to market, they need reliable and affordable supplies of seeds, fertilizers, and pesticides; they also need facilities to store their surplus crop so that they are not forced to "sell cheap and buy dear." Government policies have a strong bearing on each of these aspects of a developing country's food system.

A graphic example of such a situation was described in a recent Wall Street Journal article about agriculture and economic conditions in Africa's sub-Sahara countries[3]. It was noted that of the 30 countries classified worldwide as the "poorest of the poor," two-thirds are African. In seven of these, agricultural output is on the decline. For example, food production in Senegal is not keeping pace with population growth. In discussing the reason for the inadequate food production levels in Senegal, note was made of the fact there was little incentive for Senegal's farmers, who account for 80 percent of the country's employments, to increase their production.

Over the years, government policies in Senegal, and for that matter in most African countries, have resulted in low prices for the farmer. In part, that policy exists to ease the burden for urban dwellers, who increasingly have held the poli-

tical leverage. Under the government-run agricultural system
in Senegal, the low farm prices set by the government offer
little real incentive to grow more. Seeds and fertilizer often
are not delivered until after the rainy season. The quality of
both often is poor. Marketing is undependable. One Senegalese
farmer who grows groundnuts noted that years before the state
took over the agricultural system "everybody would try to plant
more than the next farmer. Now everyone settles for growing
the same."[3] Government policies obviously play a major role in
determining which technologies will be developed and which will
be adopted.

CONTINUING TECHNOLOGY TRANSFER CONSTRAINTS

While much progress has been made in developing truly
improved technologies which are acceptable to farmers, it is
insignificant compared to what is needed. Even for rice farmers, where a 20-year concerted effort has been made to improve
the technologies, truly superior technologies suitable to farmer adoption are available for no more than about half the rice
area of the world. While the comparable proportion of the
wheat area for which suitable wheat technologies have been
developed is higher, even here much progress is yet to be made.
Many physical-biological and socio-economic constraints remain.

Physical and Biological Constraints
A major challenge is the development of superior technology to overcome an array of physical and biological constraints on production which continue to limit the amount of
food that most Third World farmers can produce with the available resources. First, new seed varieties need to be bred with
characteristics that allow the plants to flourish under adverse
climatic conditions and can thrive with a minimum of costly
inputs. Second, we need to know how to overcome the major constraints to food production in developing countries, including
(1) insect pest and plant diseases, (2) rodents and other ver-
tebrate pests, including birds, (3) problem soils conditions,
including salinity, acidity, and zinc deficiencies, and (4)
water resource problems (whether too much or too little water).
Large geographical areas still remain where no applicable
technology exists for raising food production in economically
feasible ways. This is particularly true in semi-arid areas
like the Sahel and in some humid lowland tropical areas subject
to area flooding. Seeds of the "Green Revolution"--those that
are available today and the same ones that permitted much of
the production increases of the past few decades--flourish only
under the most favorable physical, economic and biological conditions. Plants selected on this basis are not necessarily the
most efficient when confronted with nutrient or climatic
stresses.
The challenge now is to breed new varieties that can

24

flourish under adverse climatic conditions and can thrive with a minimum or costly inputs. At the same time, we must learn how to overcome those plant diseases, pests, problem soils conditions, deficiencies and excesses of water and other production constraints which today hold millions of Third World farmers in subsistence servitude--teetering precariously on the border of abject poverty. These farmers consume almost everything they grow and thus have very little to market.

Tailored technology is what is needed--technology adapted for and tailored to local conditions and circumstances. The technology and farming techniques imported by developing nations must be adapted to site situations. Disappointment and disillusion have been the only yields harvested from many new technologies when local conditions were forced to fit the technology.

Social, Economic and Political Constraints

Certainly with regard to food production, government policies strongly influenced the relation between technological and social change in the countryside. In the case of agriculture, farmers, as mentioned earlier, will not avail themselves of new technology, no matter how accessible and productive and appropriate, unless the prices the farmer receives for his produce are set high enough to allow a reasonable profit. The ability of farmers to increase yields are also determined to a large extent by the nature for the technological innovations generated by agricultural research programs, extension programs and rural education. Government policies have a strong and sometimes decisive bearing on each of these aspects of the food system in a developing country, and positive policies produce positive results.

Land reform issues are pertinent to the successful transfer, modification, and adoption of new technologies. Taiwan and Japan underwent substantial structural changes in the late 1940's. Massive land reform programs in these two countries dissolved traditional near-feudal land tenure patterns. These reforms also expanded agricultural extension services and provided loans for peasants to buy the land they had worked as tenants all their lives. With such assistance, farmers increased the yield per hectare. Such wholesale change, however, has yet to reach the majority of developing countries.[4]

Strong local institutions are crucial to the technology transfer process. The most sophisticated scientific knowledge is of no use to Asian, African, or Latin American farmers unless local institutions can adapt that knowledge of local needs and unless extension workers can show the farmers how to put that knowledge to use on their own fields.

Institutions must be developed to enable the poor to benefit from the new technologies, or if the institutions already exist, to help them function. These institutions may be governmental, such as extension services, or they may be private, such as cooperatives or small-scale rural industries. To the extent feasible, these institutions must respond to specific needs of the poor.

The private sector vis-a-vis market mechanisms play a major role in the creation, adaptation and dissemination of technologies. However, too often consideration of institutions in development is assumed to be confined to the public sector. While it is true that in most countries public sector institutions must be relied upon to help institute new activities, and to remove production and marketing constraints, there is a growing recognition that in the long run, it is the private sector which will determine the extent of agricultural production. A very important aspect of public policies in the developing nations is the extent to which they encourage rather than constrain the private sector.

The private sector is generally more efficient in providing inputs such as chemicals and credit and in marketing the agricultural produce. The U.S. Agency for International Development recognizes the importance of the private sector and is initiating pilot programs to encourage small private agricultural enterprises in developing countries. These efforts will complement those relating to universities and government agencies.

Constraints of Trained Person Power

One of the problems of transferring technology to the less developed countries is the shortage of appropriate skills. Providing these countries with technical information does not result in technology transfer if they do not have the necessary skills to interpret it and to put it to use. The transfer of technology to developing countries, therefore, involves the transfer not only of information but also of skill, and preferably in ways that encourage the development of indigenous skills. One way to transfer technology is by person-to-person contact--bringing together the people who have the technology with the people who wish to acquire it. When technology is transferred to a foreign country, not only do you have to cross political boundaries, but also boundaries of culture, language, and custom.[5]

The most effective means of building the human resource capability of the developing countries is through formal and informal training. American universities played a critical role in the 1950's and 1960's in providing advanced training for hundreds of developing country scientists and educators. They also participated in institution development programs that provided the physical plant and faculty bases for agricultural universities around the world. A noted example is the collaborative effort of a group of universities in helping to establish 10 agricultural universities in India. Not only did hundreds of young Indians receive training in the United States, but more importantly, thousands of Indians have received their entire college education at these now well established universities. Similar, but perhaps less spectacular, successes can be noted for other educational institutions in Asia, Latin America and, to a limited extent, in Africa.

In spite of past successes in training of indigenous personnel, inadequacies of trained people remain one of the primary constraints to economic development, especially in Africa where per capita food production is actually declining. In my judgment, a massive 15-25 year effort is needed to help the countries of that continent train and educate the leaders they need to turn around their agricultural development systems.

ROOM FOR OPTIMISM

Today, I believe we have a realistic awareness of both the possibilities and pitfalls of technological change. We know that technology is far from a panacea that can overcome all the problems of underdevelopment, especially ancient patterns of social injustice. At the same time, there is widespread recognition that improved technology not only can facilitate beneficial social change but in many cases is a prerequisite for that change.

Technologies provide political leaders, extension agents, teachers and most importantly farmers, with tools for raising overall living standards and significantly those of the poor majority. Fortunately, in recent years, political leaders throughout the developing world have been showing increasing concern for raising domestic food production. Clearly, they have come to recognize that their own political and economic survival is closely tied to their ability to do a better job of feeding their own people.

I would like to conclude by emphasizing that technological change will continue to be a prerequisite for expanding food supplies, whatever a nation's policy dilemmas, social needs or ideology. By substantially increasing the supply of basic food grains, new technologies can help to hold the line on food prices, with corresponding benefits for small producers and for low-income consumers who already spend most of their limited budgets on food. By the same token, technological change is the life-blood of job creation and economic growth.

Without expecting technology to do the impossible, we must bend our efforts to develop and disseminate the kind of technology that can promote social change on behalf of the world's impoverished people--knowing that without technological advance, succeeding generations of people in the developing world will be no better off than their parents or grandparents.

REFERENCES

[1]A. W. Clausin, (Address to the Board of Governors, The World Bank, Toronto, Canada, 6 September, 1982), p. 1.

[2]R & D Mexico 2 (1982):1.

[3]Art Pine, "Sub-Sahara Countries Take First Steps to End Economic Nightmare," The Wall Street Journal (16 September 1982), p. 1, 20.

[4]Marc Kennedy, "Of Rice and Men: A Capital Development Idea," Research 16 (1982):32-33.

[5]Andrew K. Hugessen, "Transfer of Technology to the Third World" (Address to the 29th International Technical Communication Conference, Boston, 6 May, 1982), pp. 1-4.

3
Identification of Goals and Strategies in Designing Technological Change for Developing Countries

James R. Simpson

> I have met only a few government officials in de-
> veloping countries who have a clear vision of what is
> and what will be needed to modernize national agri-
> culture. The absence of such a vision is perhaps the
> single most serious impediment to developing action
> strategies for rural economic development." David
> Hopper, "Investment in Agriculture: The Essentials
> for Payoff.[1]

Three topics will be touched upon in this chapter: development goals, development strategies, and methodologies for
determining optimal country strategies. The principles and
concepts relate to the general problem of development, but
examples and orientation are to the specific case of agriculture. The major objective is relating the above topics to the
technology transfer issue with the working hypothesis being
that there is no "package" of technology which can be transferred to end-users and, in fact, that the whole concept of a
"package" has been deleterious rather than useful in designing
development strategies.

DEVELOPMENT, GOALS, AND VISION

There is considerable difference between the terms
"vision" and "goals" in development, with the principal one
being that vision is the long-term material from which goals
are set. A vision for society might be, for example, an economy based on freedom of expression, democracy, a liberal economic system, and an orientation toward improving quality of life
according to certain specified norms. It is both a qualitative
and quantitative measure. Freedom, for example, is hard to
evaluate, while literacy is easily measured. In effect, vision
is the final result, the end purpose of the development process. It is the desired accomplishments in terms of societal
improvement, or, in plain words what you would like to leave
for your kids and grandchildren. An analogy is going on a
trip; the final destination is the vision while the stops en

route are the goals. New information may cause a change in destination while a traffic accident can cause the trip to be aborted.

The problem with vision is its subjectivity and its being an ethical concept.[2] Furthermore, as new information is generated, and as the economic and social fabric of a country, and the world as a whole, changes, so must the vision. But, however difficult the definitional problem, outlining a vision is the starting place in setting development goals. Furthermore, as Joseph Schumpeter stated in his History of Economic Analysis, "...analytic effort is of necessity preceded by a preanalytic cognitive act that supplies the raw material for the analytic act...this preanalytic cognitive act will be called vision".[3]

Once a vision has been set, shorter-term goals can then be adapted to it. For example, current typical measurements of progress used in setting shorter term goals such as growth in GNP may be inconsistent with a larger vision of a quiet peaceful life. Careful analysis reveals a vision in which population growth is stabilized to be in conflict with the concept that a country's market size should be increased. In defining the vision, concepts such as equality versus equality of opportunity, and the thorny problem of income distribution (or redistribution) can be evaluated within the milieu of international trade policy, resource use, employment creation, and private versus public ownership of the factors of production.

In other words, in order to get at the "real" issues of development one must pause and ask why the issue is good or bad. In this way it is possible to carefully examine the basic premises of vision and goals. In a very pragmatic sense, vision in terms of developing country planning means calculation of what is desired at some distant point in terms of realistic "effective demand" or purchasing power, and then determining the steps to reach that desirable state.

The preceding discussion highlights an important message which is, "know where you are going." The concept, of course, is not new, and a number of agencies have attempted to bring philosophical aspects into the planning process, usually with limited success. For example, the U.S. Agency for International Development instituted their "logical framework" in the early 1970's which, although designed as an operational tool for project development rather than a guide in clarifying a country's vision, did at least attempt to point out that a specific project is part of a larger, national level set of cultural, social and economic interrelationships. Just recognizing that developing countries have more than increases in per capita income at stake is important in setting out development strategies.

Perhaps the main difficulty in setting forth a vision apart from its nebulous and ethical nature, is visualizing what a country would be like at some distant point in the future under alternative scenarios. Just thinking five years in the future at which time a specific project may be completed is no easy task; projecting out 20 years is a heroic undertaking, and 30 years is virtually beyond our ability despite the likelihood that most of us will witness those points in time. Now, this is not to say that projections have not been available, as the 1960's and 1970's witnessed a flurry of global and regional simulation exercises about resource use and misuse. Nevertheless, it is fair to generalize that projections for individual countries have not been effectively presented in a fashion which permits understandable conceptualizations of alternative situations. An approach for presentation of projections which facilitates long-term vision and associated policy setting, and a guide for shorter-term goal setting follows.

INTEGRATING VISION AND GOALS: SOME EXAMPLES

Peru
      Peru is a country on the west coast of South America with a temperate climate along the entire coastal region. Population is about 17.3 million, of which 56 percent live in the Sierra (mountain) region. The population growth is 2.8 percent annually. The literacy rate is 72 percent and PQLI (Physical Quality of Life Index, a composite index of infant mortality, life expectancy to age one, and literacy) is 65. In contrast it is 47 in Honduras and 91 in Hungary.
      About 13 percent of Peru's gross domestic product (GDP) is derived from agriculture. The growth rate of agricultural GDP was 3.7 percent in the 1960's, -1.0 from 1970-75 and 0.8 from 1975-79. The country went through an extensive land reform starting in 1969 which accounts for the negative growth in agricultural GDP during the early 1970's. The result is that the index of total food production in 1979 stood at 85, while it was 88 in 1968. The index for all agricultural production was 83 in 1979 compared with 91 in 1968. The net result is Peru's current food production situation is very serious, as population has increased 26 percent over the past decade while food production has declined.
      Moya[4] used a macro-economic demographic simulation model known as ECO-POP[5] to make 30-year projections for Peru. An unemployment rate of 20 percent was used in the base year even though official statistics estimate it at 6.5 percent. Two major projections were developed, one with an ICOR (incremental capital output ratio) of 3.8, which is just above that reported for the "normal" period of 1960-70, and another of 9.7 which is the rate for 1975-80. The projections will first be utilized to visualize the impact of population pressure on agriculture and, in the next section, the role played by technological change.
      Dividing the amount of arable land today (3.1 million hectares) by the 7.0 million people presently engaged in agricul-

31

ture yields a coefficient of 0.44 hectares per person. With a median family of six members this means 2.64 hectares of arable land per family. The amount of arable land has grown at a compound rate of about 2.4 percent annually over the past 15 years and, if that trend were to continue, it means there would be 6.3 million hectares in 30 years, just about double from current levels. Most of this would have to be in the eastern region or Selva since there is little available land for new exploitation in the Sierra, or on the coast.

The percent population in agriculture has declined significantly over the past 10 years, from 44.8 percent in 1970, to 38.1 percent in 1979. This is an annual rate of decline of 1.7 percent. If that rate were to continue, 23 percent of the population would be in the agricultural sector in 30 years. Such a fall is unlikely, so assume it only falls to 35 percent in year 30. Total population in 30 years, i.e. the year 2009, will be 39.6 million if there were a constant annual population growth rate of 2.8 percent so there would be 13.9 million people in agriculture at that time. A scenario of no increase in arable land means there would be 0.22 hectares of arable land per person, or 1.32 hectares per family. If arable land increased to 6.3 million hectares following the past trend, there would be 0.45 hectares per person or 2.7 hectares per family, about the same situation as today.

A number of scenarios could be developed using various assumptions about rural-urban migration and arable land availability. The projections demonstrate that even under the best conditions massive amounts of new arable land will be required even if heavy migration continues to the cities. The implication is that vast colonization programs will be required as there is virtually no "precapitalist" type land left to expropriate, and it is unlikely that capitalist-owned land (i.e. land on which some peasants work for wages either on a part or full time basis) will fall to further land reform.

The major point, and one which has important ramifications for development strategy, is that an analysis in which long-term projections are made can be a very useful tool in setting shorter-term targets or goals. In this case, the exercise points out the probable need for heavy emphasis on a colonization program, an approach to development that requires 15-20 years of preplanning to be successful.

Another example of vision relates to existing rural societal conditions. The dilemma, and one related to the previous example is: encourage migration to urban areas (or at least do not discourage it), or attempt to keep people on farms. This decision, which can be conscious or overt, is extremely important in planning any nation's development. At first blush the answer may seem obvious: poverty down on the farm is "better" than the more visible poverty in urban areas. Politically, it is more expedient as well as being apparently less offensive for those with short-term humanitarian interests.

But, in the longer-run, policies which attempt to prevent rural/urban migration are self-defeating in an economic sense for they are stop-gap in nature and do not solve basic problems. Again, the problem is: what do you want, economic growth which by definition virtually means industrialization of the urban sector and mechanization of the rural sector, or a stagnant GNP? Naturally, there is an optimum of some sort based on the short, intermediate, and longer-term objective vision.

Thailand

Take as another example Northeast Thailand, a vast expanse of Indochina covering about two-thirds of the country. Most of the Northeasterners are essentially subsistence-level farmers as only five percent of the area is irrigated. The Royal Thai Government recently increased the minimum level of schooling to six years and projects the level to reach grade eight in a few years. In addition, it is widely recognized by rural adults that education is the passport to a more economically rewarding life. On the other hand, the dominant religion, Buddhism, reveres peace-of-mind with strong objection to material accumulation as a dominant vision of life.

The problem, and a very serious one for people concerned with macro-level technological change, is the extent to which migration to urban areas should be encouraged in the Northeast. If the decision were made to encourage migration, then the approach probably should emphasize assistance to the educational system, and encourage the growth and consolidation of farms. Strong support could be given to a federal land bank which would provide relatively easy access to credit for farmers who wished to buy out their neighbors and thus increase the size of their own holdings. It could also emphasize intermediate-size technology at experiment stations and in the extension service. Some examples are cattle feeding in lots of 25-50 head, mechanization for rainfed agriculture, or contract growing of hogs and broilers. On the other hand, if it is decided that rural to urban migration should be discouraged, then the planning emphasis would be on improving current practices such as increasing the size of buffalo (for draft), reducing animal health problems, crop insurance for small holders, or promoting cottage industries such as silk weaving.

Usefulness of Projections

The principal worth of scenarios like the ones just presented is in assisting planners to develop realistic bounds on their vision. Peru, for example, is going to have a "minifundia" problem 30 years from now that will, in fact, only be exacerbated unless massive migration to the cities is set forth as a policy and incorporated in planning goals. Given that small farmers are usually oriented toward production of subsistence crops, the implication is that Peru will likely become a major food importer unless other steps are taken. Thailand is a major food exporter, but that situation could reverse itself now that new land is no longer available for agriculture. In brief, the long-term projections, combined with philosophical aspects mentioned earlier, can be a vital tool in setting realistic development goals, if the true interest is planning rather than crisis management.

In summary, a case has been made for recognition that since development is a philosophical problem, the first step in setting development strategy should be for planners to determine what they would like the country to be both economically and socially. Only then can goals be set as guides to meeting the vision. It is true that long-term plans can easily be abused in terms of their true purpose. It is safe to generalize, however, that some sort of a stated vision and goals statement is the foundation for development planning.

DEVELOPMENT STRATEGIES

Overall development strategies can take place from two perspectives in what could be called a "push-pull" theory of economic development. The dichotomous approaches are roughly comparable to the philosophies set forth by centrally planned economies as opposed to those which rely heavily on the free market system.

In the "push" approach government decides that it must take a major part in the production, processing and marketing processes. The size of the government sector swells as it increasingly relies on writing checks to foster economic development. In contrast to this paternalistic attitude in which entrepreneurial activity is viewed with suspicion and distrust, the "pull" approach emphasizes the use of legislation to bring about development in line with goals and some realistic vision. Planners and legislators spend their time in setting forth policies and legislation which promotes competition, yet prevents undue competition via monopolistic practices.

It can be argued that in the "push" approach government can become so heavily involved in daily management that their energy is wasted on mundane activities with the result that planning and innovation receive a low priority. In the second approach, the drawback is that government must make a decision that making a profit is desirable, and that while entrepreneurial activity is to be exalted, the creation of undue economic power will not be permitted. Clearly, self-regulation is a very difficult chore. It means fostering a development philosophy in which government views economic activity from the viewpoint of the client, i.e., consumers and businesses. It means the adoption of a positive viewpoint which translates to "what can I do to help you" rather than "how can I prevent you from doing something."

The discussion about a "push-pull" theory may seem banal

34

and far removed from daily development activities. But, it is the core of the technology transfer problem for, if the donor, whether it be the so-called developed countries, LDC government, or agricultural extension service views itself as having a "package" to sell and promote, it puts the donor in the role of a "pusher."

Far more important is making a decision to make a decision to view the planning process, and the technology process associated with it, from the client's viewpoint with that decision must be a recognition that every situation is different. This means, of course, subscribing to an "everything depends on everything" school and with it a belief there is no one development or technology package. Since there is no package, it means that people with vision are required who are both change agents in the development process, as well as being technicians.

THE IMPACT OF TECHNOLOGICAL CHANGE: PART TWO OF THE PERU EXAMPLE

Low incremental capital to output ratios (ICOR) are desirable for they mean that relatively small amounts of <u>existing</u> capital are being used for each unit of output. In contrast, high coefficients mean that large amounts of <u>existing</u> capital are required per unit of output. In the case of Peru, due to the political instability of the past several years, each unit of capital is now yielding a relatively small amount of output. Following is a summary of GNP per capita with two different ICOR's and constant population growth:

Assumption	ICOR[c] = 3.8		ICOR = 9.7	
	Constant		Constant	
PGR[a]/	2.8%	2.8%	2.8%	2.8%
TCR[b]/	1.5%	1.5 to 2.5%	2.5%	1.5 to 2.5%
Year	----------------Constant $US----------------			
0	672	672	672	672
30	1,520	2,176	1,129	1,596

[a]PGR = Population growth rate.
[b]TCR = Technological change rate.
[c]ICOR = Incremental capital-output ratio.

In contrast to the above predictions, when population growth rate is allowed to decline from 2.8 to 1.5 percent in year 30, and technology increases from 1.5 to 2.5 percent, then

GNP per capita increases to $2,914 with an ICOR of 3.8, and $1,955 with an ICOR of 9.7. The conclusion is that technology is very important in improving per capita income, especially when it is combined with efficient use of existing capital stock.

Finally, the projections show the great improvement possible from combining improved technological change with efficient use of capital stock and a declining population growth rate. The range for year 30 in per capita income is quite large, from $1,129 (only a 68 percent increase) under the worse case (constant population and technological growth rates, and an ICOR of 9.7) up to $2,914 (a 334 percent increase) when ICOR drops back to 3.8, population growth rate declines to 1.5 percent, and the rate of technological change increases to 2.5 percent. Again, the assumptions can be discussed and changed, but the results are clear: economic benefits are clearly related to technological change, capital use and population growth.

METHODS IN THE TECHNOLOGICAL CHANGE PROCESS

The initial discussion of methodologies, i.e. philosophies about the change process, are useful in terms of orientations. The purpose of this section is suggesting some concepts and aspects which are most directly applicable to formulation of development plans about actual or current problems, i.e. ones of immediate concern. We begin by reiterating again the working hypothesis--that there is no one technological package.

An interesting point of departure is recognition that a virtual revolution has taken place in the development of planning techniques and the means for applying them to developing country situations. Following are some of the more common evaluative techniques available to economists. The list is not meant to be exhaustive; rather the purpose is developing an appreciation for the complexity involved in selecting the appropriate tool which is most cost-effective for a specified accuracy level. Some of the tools have been available for many years but are just now being used to a greater extent due to the advent of computers. The techniques are: partial and complete budgeting and related sensitivity analysis, correlation, cross-sectional analysis, deflating and indexing, econometric models, PERT/CPM, input-output analysis and multipliers, linear programming, simple and multiple linear and non-linear regression, project analysis, quadratic programming, simulation, and transportation models.

The above list is rather impressive for it indicates the level of sophistication required first in choosing the appropriate tool (which assumes that the project investigation has been adequately specified prior to determining the technique or tool), and second in its cost-effective application. The point is, we are being hurtled forth at a geometric rate into use of

sophisticated tools and procedures, and those countries which
do not adopt them in the planning process are going to be seriously disadvantaged. In effect, where aid programs previously
viewed advanced education only as a necessary subcomponent of
the development process, there will have to be increased recognition that technological change is, largely, a people and
knowledge process.

It is fair to state that, in the past, technological
change has been viewed in a physical context. Appropriate
technology has meant small-scale tractors for small plots of
land; better combinations of plants in a multiple cropping context; a cheap solar oven for areas with little available cook-
ing fuel; or increased calf crop through better nutrition.[7]
But, just as economists have generally relegated people problems to the last page of their report with a warning that they
must be taken into account, so have people generally been relegated to footnotes in articles about technological change. The
bottom line is that just as there is a serious missing link
between farmers, extension and agricultural researchers, so is
there a missing link to technology adoption in development.

GETTING AGRICULTURE MOVING

About 15 years ago Arthur Mosher published a little book
called Getting Agriculture Moving.[8] In it, he concentrated
heavily on portraying agriculture from the viewpoint of
clients, i.e, farmers themselves. Since then a large number of
books and articles have appeared about development strategies
and methods, most of them biased heavily toward "The" prescrip-
tion to obtain development. A few examples are instructive.

A recent widely used book in agriculture is Wortman and
Cummings, To Feed This World: The Challenge and the Strategy,[9]
a marvelous work which spans the entire world. But, careful
reading reveals that the human element is virtually left out,
as the entire focus is on biological processes with planning
discussed in terms of campaigns, programs, and support. Their
approach is essentially one of crop yield improvement.

The opposite extreme to Wortman and Cummings, at least in
the western world, is de Janvry's 1981 book The Agrarian Question and Reformism in Latin America.[10] This work, completely
based on marxist principles, has almost nothing to say about
the biological process, with the conclusion drawn that integrated rural development, "...focus(es) on the creation of a
pampered minority fringe of petty bourgeois who serve as an
essential buffer between bourgeoisie and semiproletarianized
and landless peasants; their economic interests are tied to
those of the bourgeoisie while their ideological identifica-
tions lies with the bulk of peasants, who recognize them as
patrons".[11] His solution for Latin America is:

......in agriculture, the political program should
thus be one of fomenting an alliance between work-

37

ers and those segments of the peasantry that can
be mobilized for this purpose...and it is important to create a state of consciousness among
peasants that allows them to see beyond petty
bourgeois demands. For this purpose, redistribu-
tive land reforms and rural development programs
conducted by (and not for) peasants can be useful
departing points in a struggle for democracy and
articulation.[12]

An intermediate approach to those above, is the farming
systems research and extension program (FSR/E) developed in
Guatemala. As with any method and methodology which has
matured in a short time, it has strong critics as well as
adherents. But, it does have one main feature which has wide
applicability to all sectors of an LDC economy--it is a bottom-
up approach which begins with a multidisciplinary quick survey
of an area to understand the situation and outline systems. A
unique feature is that a significant amount of the experimental
and extension-oriented work is to be done on the farm rather
than at experiment stations. Farmers thus become involved in
the process both in the evaluative and extension stages.

The trouble with this approach is, of course, that the
change agents, donors, government institutions, and other
related parties must be convinced that a "pull" approach is
better than a "push" approach. It also starts with a problem-
solving rather than descriptive orientation, and one in which
the change agents (for lack of a better term even though they
may be agronomists, plant breeders, entomologists or anthropologists) are determined to look at the situation in a holistic
manner from the farmer's viewpoint. This also implies, for
instance, that a plant breeder has to look at corn as a commodity and a total production problem, rather than as one of seed
type "A" versus seed type "B".

The concept just described can be termed "in-place" technology development, and FSR/E people are the pioneers in it.[13]
The importance of this approach cannot be stressed too heavily,
for it requires an entirely new mode of thinking--and that is a
very difficult proposition. It is probably as difficult to
absorb as the concept of vision and goals in integrated development planning--and rejecting the concept of a package of
technology.

SUMMARY AND CONCLUSIONS

A rather wide variety of topics have been covered in an
effort to demonstrate several principles:

1) The starting point for effective planning is 20 or 30
   year projections of economic-demographic variables.

2) The next step is articulating a long-term vision of

society's desirable shape in economic-subjective terms tempered by the realism derived from the economic-demographic projections.

3) After the vision is established, key areas for concentration of resources can be identified.

4) Once the general directions are set out, short-term goals (e.g., 5 year) can be developed.

5) Specialists and experts tend to give recommendations about the planning and development process from their own point of view. The vision and goals are best set by a multidisciplinary team of consumer interest groups, government, and specialists.

6) The vision and goals of government are radically different from those of individuals and firms. The conflicts are natural. They simply need to be recognized and then resolved.

7) A "pull" type of approach is recommended over the "push" approach.

8) There is no overall package of technology for development planning as each situation is radically different.

9) There is no package of technology, even for very specific situations such as pest control in corn.

10) Technology and the procedures used for it, and in its adoption in the planning process, are increasingly becoming more complex. This provides both an opportunity and a challenge in development planning.

11) Technological change is much more of a human problem than previously thought. Consequently, much more orientation should be placed on human capital development, especially in fostering imagination and innovation. Economic development takes place because people want it to happen.

12) The approach taken by farming systems research and extension has wide applicability for agriculture. Much more attention needs to be given to it. Furthermore, the concepts can be transferred to other sectors of the economy.

Overall, the most important factor is recognition that each development situation is different, that the technological change process is primarily a human one, and that development planning should start with a long-term view from which short-term goals or targets are set. Only with a clear vision of what is needed and desirable can "development" take place.

REFERENCES

[1]W. David Hopper, "Investment in Agriculture: The Essentials for Payoff," Strategy for the Conquest of Hunger: Proceedings of a Symposium (New York: The Rockefeller Foundation, 1968).

[2]James R. Simpson., "Ethical Issues and International Development: New Challenges for the Agricultural Economist," The Rural Challenge: Contributed Papers at the 17th International Conference of Agricultural Economists, IAAE Occasional Paper no. 2 (Jakarta, 1981), pp. 319-324.

[3]Joseph Schumpeter, History of Economic Analysis (New York: Oxford University Press, 1954).

[4]Joaquin Moya, "Evaluation of the Peruvian Agricultural Economy: Recent Historical Aspects and Estimates for the Future" (Food and Resource Economics Department, University of Florida, June 1982).

[5]Ray Billingsley, and Douglass Norvell, ECO-POP: A Macro Economic-Demographic Simulation Model (College Station: Texas A&M University, 1969); James R. Simpson and Ray V. Billingsley, "Application of a Macro-Economic Demographic Simulation Model to Planning in Paraguay," International Journal of Agricultural Affairs 8 (1974):78-88.

[6]Alain de Janvry, The Agrarian Question and Reformism in Latin America (Baltimore: The Johns Hopkins University Press, 1981).

[7]Leo Polopolus, "New Technology in Agriculture and Its Potential Impact" (Paper presented at the 1981 International Annual Meeting, Animal Health Institute, Ashville, North Carolina, September 25, 1981).

[8]Arthur Mosher, Getting Agriculture Moving (New York: Praeger, 1966).

[9]Sterling Wortman and Ralph W. Cummings, Jr., To Feed This World: The Challenge and the Strategy (Baltimore: The Johns Hopkins University Press, 1978).

[10]Alain de Janvry, The Agrarian Question and Reformism in Latin America (Baltimore: The Johns Hopkins University Press, 1981).

[11]Ibid, p. 253.

[12]Ibid, p. 268.

[13]W. W. Shaner, P. F. Philipp, and W. R. Schmehl, Farming Systems Research and Development: Guidelines for Developing Countries (Boulder, Colorado: Westview Press, 1982).

4
Institutional Linkages in Development Projects with Special Reference to Marine Fisheries

James A. Storer

This chapter will be concerned with some of the special macro aspects of development process as they apply to marine fisheries, particularly institutional elements and linkages at the macro level. It may well be that a concern with marine fisheries is somewhat out of the mainstream of overall economic development. Nonetheless, the importance of fish as a protein source is established--even though this importance looms greater to many less developed countries (LDC's) than it does to us in the United States. In Southeast Asia, for instance, fish amounts to well over half of the total animal protein diet.[1]

WHY FISH ARE DIFFERENT

The rationale and need, however, for looking at fisheries development as something different and requiring special handling rests upon two elements. First, and most important, fish living in the sea display varying degrees of mobility and are considered as common property. This last feature means that no one fisherman has any incentive to maintain the productivity over time of this renewable resource, for what he does not capture someone else will. When fishing levels get to be heavy, therefore, it is necessary to superimpose some system of management that restores the fishing effort to a level that the resources can withstand over time

A second element about marine fisheries is a recent phenomenon and essentially stems from the first physical condition mentioned. This is the virtually worldwide establishment of 200 mile zones of national fisheries jurisdiction, a development which spread rapidly after the U.S. took the unilateral step through legislative action in 1976. By the end of 1979, 99 percent of the world's fish catches were contained within these zones.[2] The pressures to take this step were several, but most significantly they were related to the greatly intensified fishing effort that took place after World War II that had seriously threatened the economic availability of certain stocks and had led to severe declines in the catch of a number of the most valuable species. Much of this greater effort came from distant water fleets, (Russian, Japanese, Polish, and Korean), that fished on relatively shallow continental shelfs and banks of other coastal states, including the U.S.

MANAGING THE RESOURCE

Though efforts had been made through a number of international fishery commissions and bilateral agreements to control coastal fishing effort and attain some degree of management of fishery stocks, these efforts were consistently "too little and too late." Putting up a fence, therefore, at the 200-mile point would keep foreigners out except under terms specified by the coastal state and would also presumably give the coastal state some ability to manage the fishing activities of its own citizens.

A number of developing states perceived that in controlling the access to their waters by foreign fishing fleets they could earn foreign currency by selling license fees and could also require participation by foreigners in shore-side activities and other joint ventures that would develop the coastal state's own fishing industry. Creation of the 200-mile fishery zone opens up the possibility of excellent development opportunities for national governments. But these opportunities may never become a reality, or may be fleeting indeed, if they are not developed in the context of an adequate system of resource management.[3]

So far, with few exceptions, coastal developing states have not been able to grasp effectively the opportunity. At times there has been an excessive exploitation as in the Gulf of Thailand, where Thailand mobilized a trawler fleet that rapidly overfished the stocks with disastrous results. In other cases, coastal states, through a lack of administrative or enforcement capability, have let foreign fleets operate in their zones without adequate controls from a conservation or an economic viewpoint.

These problems have been stated by a working party of eminent fishery scientists brought together in 1980 by FAO. Noted that there is a finite opportunity - a window in time - in which the world community and the coastal states can work together to place new national fisheries on a solid groundwork of management.[4] Delay is dangerous. If action is not taken soon --within five years or so--the opportunity to avoid painful mistakes will be lost, and correction in the future will be costly, not only in money, but in terms of social dislocation. The FAO group further concluded that fisheries projects which are specific and finite and have a return that can be qualified will not take adequately into account the need for management.

The significance of all of this for fisheries development is to emphasize the vital importance of the institutional mechanisms, capabilities and linkages that must be present if

44

marine fishery production and economic viability are to be
attained and maintained.

INSTITUTIONAL LINKAGES

Because of its importance and uniqueness, some further
discussion of these institutional needs and their relationships
or linkages is warranted. At the outset, it is recognized that
fisheries planning for development is primarily a national matter. It is also true that the management of a country's fishery resources within its fishery zone is primarily a national
concern. Indeed, most fishery zone legislation or proclama-
tions virtually assert sovereignty over the fishery resources
which is contrary to the sense of trusteeship and international
responsibility that is to be found in the text of the Law of
the Sea Treaty.

Although fisheries management is usually a function of a
national government, there are strong local or regional interests to be considered. Indeed, it is often at the local level
that competing interests for the same fishery resources are
encountered. Furthermore, the necessity of adequate communication from the national government down to the lowest level of
community participation is requisite if fisheries management is
to be supported and successful.

Despite this challenge, the capability of most LDC governments to respond is limited indeed. Fisheries administrators
are normally forced to work with a very small and underquali-
fied staff. Most frequently, the training of the staff is confined to biological experience, if that, and other necessary
disciplines are rarely to be found. This situation merely
reflects, however, the state of affairs found even in developed
fishing nations where professional experience and training in
other than biology is still notably lacking.[5] Even in the
U.S. today, a genuine inter-disciplinary approach to fisheries
development management has only begun to be taken. Furthermore, the capability for professional assistance from universities in LDC's is scant indeed. While there has been a tradition of sending biologists abroad, there has been little done
to increase the availability of trained fisheries economists,
planners, or others in the social sciences. It also appears
that fisheries officials do not normally rate very high in the
overall administrative hierarchy and often remain effectively
out of communication with the power center in national governments, including planning offices.[6]

Despite the importance of the national government, and no
matter how adequate or inadequate it may be, the national government must inevitably look to other levels in order to
achieve its development and management objectives. And in this
context I am not referring only to the need for outside assistance from either other national governments or international
organizations, though it is addressed.

45

INTERNATIONAL EFFORTS

The need for involvement with other national governments, international organizations or regional commissions stems from the nature of the resource and the fact that fish swim freely across national boundaries. Thus, while the U.S. with its long coastline is relatively free of need to consider other countries in its fishery management measures (with the notable exception of Canada) most countries, such as those in Africa bordering the eastern Atlantic, can only successfully develop and manage their own fisheries within the context of an effective regional mechanism.[7]

Fortunately, there is something of a network of regional commissions around the world, a number of them having been sponsored by FAO. Equally fortunate, these FAO commissions have recently been concerned with both regional development as well as management in their particular areas. Thus, there is an opportunity within the same forum to consider the interaction of increases in fishery effort and investment along with the concomitant impact upon the shared fishery stocks. The unfortunate aspect is that the authority of these commissions in the field of management is necessarily very limited and they are as yet quite untested in trying to achieve any semblance of a management system in consort with the coastal states concerned. A central problem in any regional management system is that each national government must surrender some of its authority, an action that most states are loathe to accept.

It should also be noted that just at this time when the role of the FAO regional commission needs to be strengthened, the funding for the related regional development projects is being reduced or eliminated because of the budget limitations of UNDP. Thus, despite the recognized need for effective regional organizations to play a role in the overall development and management of fishery resources, it is not clear how much can be achieved.

I have already referred to the role of the FAO and UNDP as international mechanisms. Perhaps a brief further discussion is appropriate regarding this highest link in the institutional chain that leads to the beach. FAO's Department of Fisheries is the only worldwide fisheries organization that has the expertise and experience to address the whole range of scientific and technical problems that confront its member countries. It is also the only source for data on world fish catches and trade, the importance of which alone justifies the existence of the Department. With the establishment of fisheries zones throughout the world, FAO has mounted an ambitious program to try to meet the pressing needs of LDC's, emphasizing both the regional aspects and the necessary relationship between management and development. Unfortunately, bureaucratic problems, a current lack of leadership, as well as budget difficulties have severely handicapped FAO's ability to respond.

DONOR GOVERNMENTS AND FISHERIES DEVELOPMENT

The role of donor governments, operating bilaterally, in providing assistance to LDC's requires some comment with respect to fisheries. Since governments, notably Norway, Sweden and to some extent Canada, have given a considerable priority to fisheries development and have operated programs that had continuity to them and were linked closely or were an intensive part of broader regional projects undertaken by FAO/UNDP. This maximized the impact and served to reduce some of the duplication in staffing and other costs.

The U.S. program of fisheries assistance operated by AID has recently been the subject of a lengthy study by the Ocean Policy Committee of the National Academy of Sciences.[8] There are, of course, a number of shortcomings to the AID fisheries program, not the least of which is the relatively small emphasis fisheries has received and the reluctance of AID to consider the broader and unique aspects of fisheries development and management. This is all the more regrettable since the United States, by its own unilateral action, was responsible for setting the pattern for worldwide extended fisheries zones.

This failure of AID is brought about in part by the lack of a strong fisheries voice in AID headquarters and general lack of fisheries expertise in the field missions. However, one should also note with approval that AID has for a number of years provided support to two U.S. universities active in fisheries development. One, of course, is Auburn University with its program in aquaculture which has been particularly successful in establishing and maintaining linkages to LDC's not only through its own extensive field experience but also in its training activities carried out in Auburn. The other university is Rhode Island which has tended to concentrate on marine and artisanal fisheries and has tried to present a genuinely inter-disciplinary approach.

More recently, AID has embarked on funding a new kind of international effort--that is a non-governmental organization called ICLARM, the International Center for Living Aquatic Resources Management.[9] This entity, located in Manila, was conceived, and still largely supported, by the Rockefeller Foundation. ICLARM is not a "bricks and mortar" institution. Rather it is a small group of scientists and experts, strongly inter-disciplinary, who work on a variety of research projects in both aquaculture and marine fisheries. They have established close and effective working relationships and links with universities, fisheries administrators and other organizations in Southeast Asia. Because they are non-governmental, they have a greater flexibility and freedom and have been able to respond quickly and effectively to the needs of LDC's. While ICLARM is in no way a substitute for FAO or other international mechanisms, its unique approach, which is only as good as its staff, presents one of the brighter spots in the current range of institutional mechanisms to deal with fisheries development and management.

CONCLUSION

The previous paragraphs have outlined the multiplicity of levels that must be involved in the planning and execution of fisheries development and management. It may appear that the complexity of the process, and the problems, as well as the limitations of expertise and funding conspire to make this a most discouraging field with which to deal. Yet, some of the most notable rates of growth of fish catches are occurring in LDC's--at times through the cooperation with the private sector in developed countries through a variety of joint venture activities. Furthermore, there is added evidence that the LDC's themselves see the dual necessity for development and management and are prepared to deal with the range of problems, locally, nationally, regionally and internationally.

I hope that the U.S. can fulfill a proper role in assisting LDC's to make a more effective and lasting utilization of their fisheries resources. This role would involve the cooperation of a number of governmental organizations, notably AID, but also including the good work of the Peace Corps as well as a number of universities and private institutions.

Nothing could be more helpful or useful in this important matter, than if AID would clearly indicate its commitment to fisheries development and management and embark on a program which, although necessarily limited in funds, indicated an awareness of the nature of the problems and a willingness to support projects that would work towards a long-term solution.

REFERENCES

[1]Food and Agriculture Organization, World Fisheries and the Law of the Sea, (Rome: FAO, 1981).

[2]Roy Jackson, Extended National Fisheries Jurisdiction, Washington Sea Grant Publication (Seattle: University of Washington, 1982).

[3]Sea Grant Institute, Toward Future Fisheries Management: Some New Concepts for the 1980's (Madison: University of Wisconsin, 1981).

[4]"Report of the ACMRR Working Party on the Scientific Basis of Determining Management Measures," Food and Agriculture Organization of the United Nations, Fisheries Report No. 236 (Rome, 1980).

[5]Kenneth Craib and Warren Ketler, eds., Resource Development Associates, Collaborative Research in the Developing Coun-

tries - A Priority Planning Approach (Prepared for the U.S. Agency for International Development and the Joint Research Committee, Board for International Food and Agricultural Development, Washington, D.C., 1978).

[6]Marine Technical Assistance Group, Ocean Policy Committee, National Research Council, International Cooperation in Marine Technology, Science and Fisheries (Proceedings of a Workshop, Scripps Institution of Oceanography, La Jolla, California, January 18-22, 1981).

[7]Edward Miles, "On the Roles of International Organizations in the New Ocean Regime," (Prepared for the 14th Annual Conference of the Law of the Sea Institute, Kiel, 1980).

[8]Marine Technical Assistance Group, Ocean Policy Committee, National Research Council, An Evaluation of Fishery and Aquaculture Programs of the Agency for International Development (Washington: National Academy Press, 1982).

[9]International Center for Living Aquatic Resources Management, ICLARM Report 1981 (Manila: ICLARM, 1982).

5
Establishing Appropriate Linkages from Development Projects to the Central Government

W. W. Shaner

A development project may be considered as some well-defined activity, usually an investment but not necessarily so, that the government of a developing country considers as contributing significantly to the economic growth or the general welfare of its people. Since the central government customarily plays the principal role in the country's development program (with varying degrees of regional and private participation), it will normally exercise considerable control over development projects. This control shows up through government investment, participation in ventures with local and international groups, authorization of privileges or imposition of restrictions on specific ventures, or establishment of policies and institutions to facilitate investment and greater productivity throughout the country. The central thought of this definition is the central government's involvement in a concerted effort aimed at the country's development. For other definitions and descriptions of development projects, see Baum[1] and King[2].

Historically, the lack of well-conceived projects has too often been the primary weak point in the planning process. For example, Waterston has said that the "weakness in most developing countries is not the lack of an elegantly integrated comprehensive plan based on economic potentialities but the lack of well-planned individual projects that can really be carried out."[3] He goes on to say that after 18 months in one country the planners found themselves in the embarrassing position of conceding that the principal deficiency was the small number of specific investments available to implement their plan.

In pursuing this theme and relating it to the transfer of food production technology to help developing nations attain their objectives, this chapter (1) examines the principal steps in project planning and activities, (2) turns briefly to the means for transferring technology, and (3) closes with some issues for your consideration.

Throughout this paper, I have drawn heavily on my experiences while working with the Imperial Ethiopian Government's Planning Commission, as a consultant to various water resources groups in Peru, and as one of the principal authors of a recently completed study of farming systems research (FSR) methodology.

STEPS IN PROJECT PLANNING AT THE CENTRAL GOVERNMENT LEVEL

Project planning at the central government level can be broken down into project identification, design, analysis, implementation, and evaluation. Each of these topics is discussed below. Note that this breakdown is similar, but not quite the same as the breakdown by Baum in which he defines the project cycle as identification, preparation, appraisal, negotiations, and supervision.[4]

Project Identification

Project identification, an initial activity in project planning, is one of the outputs of macroeconomic planning whereby projects along with policies and institution building are the principal means central governments employ for implementing development plans. Earlier, the macroeconomic planners will normally have considered the country's national and regional objectives, resources, opportunities, and problems. Typically, economic growth, more equal distribution of income, and employment have been high on developing countries' lists of priorities.[5] Now, with the concern over food shortages and high rates of population growth, more attention is being placed on self-sufficiency in food production.[6] Improving the productivity of small-scale farmers is seen by some as a way of simultaneously increasing national food production, improving the distribution of income, and addressing unemployment and rural-to-urban migration.[7]

Once the planners and those brought into the planning process settle on these broad issues, their next step is to decide on a development strategy. Candidate strategies include for example, an emphasis on industry, agriculture, exports, import substitution, infrastructure, balanced growth,[8] and unbalanced growth.[9] Which of these, or other strategies, will be selected for the plan depends in part on the Interplay of a country's previous activities and customs and on the influence of those developing the plan. For instance, Ethiopia's Third Five-Year Development Plan (1963-1968) had the imprint of the World Bank because of the assistance the government received from that institution, whereas the previous development plan was influenced by the several Yugoslav advisers assisting the planning commission at that time.

Sector, resource, and related studies, meetings, and other activities provide the planning group with information on ways to implement the selected strategy. And details of the plan, such as sector targets, investment requirements, and sources of funding are reconciled through what Lewis calls the "arithmetic of planning."[10]

In the process of developing the foregoing information and in making the needed calculations, the planning team will uncover investment and other project opportunities. These opportunities may come about from sources such as sector studies, identification of shortages, import substitution and export possibilities, solicitations from industry, input-output analyses, and so on. A particularly interesting approach to project identification was that taken by Lewis in which he evaluated opportunities for imports, exports, and production for local consumption on the basis of weight-to-value and labor inputs.[11]

Throughout this project identification process, the planners need to relate the ways in which individual projects contribute to the development strategy and ultimately to the country's development objectives. For example, if self sufficiency in food production is an objective, how would a project contribute to this end? If regional employment is an objective, how many jobs does the project generate? Or, if foreign exchange shortages are constraining the country's rate of growth, what will be the project's impact on the foreign exchange balance?

Project Design

Another way planners can contribute to a country's objectives and take advantage of its resources and opportunities is through project design. For example, projects that contribute to food production can also contribute to employment and improvements in income distribution through selection of labor intensive means of construction and operation. Shadow pricing is commonly used to arrive at efficient resource use from the national perspective,[12] while other objectives such as employment and income distribution can be handled as separate measures of a project's value.[13] This linkage between projects and central planning is crucial to project planning in developing countries because projects are frequently more effective in attaining government objectives than other measures, e.g., progressive income taxation.

Besides this linkage, another aspect of project design to consider is its complexity. Designing projects that will be significant for society tends to be complex partly because the designers need to consider why similar projects have failed in the past or were not undertaken at all. To allow for this complexity, planners need to integrate a variety of disciplines and consider a range of alternatives. I know of no examples in the development field where such complexity was adequately handled through systems analysis techniques such as linear programming or simulation. I am not suggesting that parts of the project design cannot be optimized using optimization procedures. Often they can. What I am saying is that overall project design does not easily lend itself to global optimization.

Rather than seek perfection in project design, I prefer to search for rationality of design. By that I mean a design in

which the project's components have a reasonable chance of being accomplished as planned and do not incorporate obviously
gross inefficiencies. Winkelmann and Moscardi of CIMMYT call
this a "nonperfectabilitarian" approach, which simply means
producing a better solution for a particular problem rather
than seeking the best solution.[14] Not only are "best" solutions difficult to obtain, they are also time consuming; and
time has its opportunity costs too.

Thus, in designing an agricultural project, for example,
planners would be urged to identify farming enterprises and
practices that farmers find acceptable and for which markets
for the output, incentives, and financial resources are adequate. In producing such a practical design, the planners may
need to call on those with skills in disciplines such as agro-
nomy, animal production, agricultural economics and engineering, and rural sociology.

Project Analysis

Provided the project is adequately designed, the next
question is whether the project is in the economic and financial interests of both the participants and the central government.

I prefer to begin the analysis by applying market prices
to the project's direct inputs and outputs, assuming the
private sector has some involvement. Such an analysis calls
for a judgment about what is an adequate incentive for the
private participants. If they are private businesses, the
measure of incentive might be some minimum attractive rate of
return; and if the participants are individual farmers, the
measure might be some level of increased income or output from
the farmers' land, labor, capital, and management--depending on
how these factors are controlled by the farmers.

If the private participants have sufficient incentive, the
next question is whether their financial resources are sufficient to meet cash expenditures, including any debt servicing.
If so, the project is financially feasible from the private
point of view; if not, the project as designed fails this
financial feasibility test. Closely allied with this question
of financial feasibility is the attitude of small-scale farmers
toward cash expenditures and borrowing. According to Zulberti
et al., many farmers when choosing among alternatives favor
those with low cash requirements.[15] Thus, even though a project might appear financially feasible because the participants' cash resources are adequate, it might in fact not be
because of the participants' reluctance to increase their cash
expenditures.

Whenever a project fails the economic or financial test
from the private viewpoint, the government is left with the
alternatives of abandoning or redesigning the project or alternating conditions in the private sector. This decision depends
on estimates of the project's economic attractiveness from the

national perspective--sometimes called the social benefit-cost analysis. This analysis considers a project's benefits and costs from the national viewpoint.

Three factors distinguish this type of analysis from that undertaken from the private viewpoint. First, shadow prices are used in place of market prices. Shadow prices represent the value, sometimes called the opportunity cost, of inputs and outputs to society. For example, if unskilled labor is in surplus supply, its opportunity cost to the economy could be considerably less than its market price. Estimating shadow prices is a proper function of the central government whereby the country's present and future resource requirements and availabilities are best understood.

Second, since indirect effects, by definition, apply to those outside the project, they are usually of little concern to the project's private participants. In contrast, by virtue of its nationwide responsibilities, the central government is rightly concerned about what happens to those outside the project. Yet, properly taking indirect effects into account can be difficult because of their pervasiveness and the difficulty of obtaining reliable data.

As a result, some analysts advise planners to ignore all but the most obvious of the indirect effects, e.g., use of large surplus capacities and serious degradation of the environment. This approach to indirect affects allows planners to (1) concentrate their efforts on a project's direct effects where data are firmer,[16] and (2) reduce the opportunity for some to unduly inflate a project's benefits by ignoring its indirect costs or competing projects' indirect benefits.[17]

The third adjustment concerns the project's contribution to those national objectives not easily valued in monetary terms. These objectives include employment, regional development, improved income distribution, national security, and the like. Perhaps the simplest approach for handling this adjustment is to quantify in physical terms the project's contribution to these objectives and then to present the results to the government's decision makers as additional information about the project.

If, on balance, the project seems in the national interest, the planners must still test for the project's impact on the government's financial position. Anticipated flows of local and foreign currencies associated with the project serve as a base for estimating initial monetary and debt servicing requirements. Logically, projects judged as being highly attractive to society should be financially viable because of the government's command over local and foreign currencies. In practice, however, exercising such command is constrained by existing policies and practices and by competing interests.

In my experience, this sequential approach to project analysis has proved highly manageable, but other approaches are possible. More complex approaches such as those by Little and

Mirrlees,[18] Squire and van der Tak,[19] and UNIDO[20] have the advantage of integrating economic efficiency in resource use, other national objectives, and certain elements of development theory--e.g., the marginal propensities to consume according to income groups. The Squire and van der Tak approach, for example, includes weights for the government's preferences for economic growth and income distribution. Consequently, their results provide an integrated value for ranking projects. Experiments with similar approaches have been conducted by several analysts such as Bruce[21], Lal,[22] and Little and Tipping.[23] Beyond this, it is not clear how widely such approaches have been applied by planning units in the developing countries, but their applications have been limited. Gittinger writes that the World Bank recognizes the significance of the concepts proposed in these more complex methods, but generally stops short of employing their analytical procedures.[24]

Project Implementation

During project identification, design, and analysis, the planners will sometimes have either general or specific knowledge about how a project is to be implemented. Implementation requires knowing who the investors and operators of the project will be, requirements for institutional and policy support, and means for funding the project.

An often critical issue facing the planners will be who will own and operate project facilities. Sometimes this decision is constrained by existing policy, such as Mexico's requirement that mining activities be controlled by Mexico's and Spain's policy of government investment in strategic industries whenever the private sector fails to meet the economy's needs. At other times the options are more open. For instance, during the late 1960's, the Ethiopian government explored a number of cooperative ventures involving various combinations of government and private investments and means of operation. In this work, we sometimes encountered what appeared as serious offers by national and private investors but were really thinly disguised tactics for selling equipment. Probing the offer and the investor's background was usually sufficient to identify invalid investment proposals.

For government controlled activities, planners need to weigh the relative merits of having ministries or government corporations manage project operations. Where a project produces revenues, as with power generation, or where freedom from existing policies is advantageous, government corporations can be a useful organizational form. For example, because it was created as a government corporation, Guatemala's Agricultural Science and Technology Institute has been able to exercise greater freedom in budgeting, staffing, and related matters.[25] However, such corporations have their problems when revenues fail to match expenses, when the corporation no longer adequately serves the national interest, or when its role is unclear.

As an example of the latter, a group of us were once asked to study the problems of a national shipping line that was unable to meet expenses, thereby making significant demands on the national treasury. The major cause of the losses was the high insurance rates being paid because of the line's subsidized rates for hauling oil cake. The shipping line's manager continued authorizing high insurance rates being paid because the line continued hauling oil cake, which had caused several fires. The line's manager subsidized oil cake rates because he felt that he was supporting government policy to earn foreign exchange and to assist the oilseed industry. Eventually, these losses brought him into conflict with the Ministry of Finance, which had to cover these losses. Such a situation called for the central government to decide where its priorities lie.

As concerns policy and institutional support, projects are often required to function within existing policies and with existing institutions. Otherwise, frequent policy changes to accommodate individual projects may confuse potential investors and could be unfair to past investors. Moreover, while institutional development is a legitimate goal of a central government, such development takes time and resources. Consequently, project staffing and organizational requirements are often accomplished incrementally rather than through major changes instituted abruptly. Of course, very large projects--especially those that develop over time--may justify policy and institutional changes.

Another part of project implementation is the securing of adequate funds for investment and operations. Because access to funds influences a project's scale, timing, and mode of operation, those designing a project will normally want to have a reasonably clear understanding of funding alternatives. Other financial aspects include identifying the time and size of monetary requirements, effective interest rates for loans, debt servicing, and financial control of operations.

International lending institutions often press for the beneficiaries of a project to pay as much of a project's cost as possible. As a minimum, production-oriented projects should provide sufficient returns to cover operation and maintenance costs. Otherwise, a government might eventually find itself burdened with projects for which its revenues are insufficient for operation and maintenance of facilities.

In one instance, a USAID mission exerted pressure on a government to recover more of a project's costs through higher water-use charges. However, USAID's efforts met with considerable resistance because of the country's tradition for pricing water at nominal rates and the inability of the administering agency to change water prices. Of course, charging less than full project costs is a means for redistributing income in favor of project beneficiaries.

Another financial aspect that affects organizations providing technical assistance to developing countries is USAID's policy favoring host country contracts. The purpose of such contracts, involving loans from the U.S. government to the host country, is to reduce USAID's involvement in contract negotiation and administration by having these activities picked up by the contractor. While relieving USAID of considerable contractual detail, the effect of this policy has left a number of contractors dissatisfied. Reasons for this dissatisfaction include: (1) the contractors were not adequately knowledgeable about the host country's contracting laws and regulations and could not easily protect themselves during contract negotiations, (2) performance bonds and potential arbitration of disputes in the host country placed the contractors in a vulnerable position, and (3) attention to such details greatly reduced the time available for performing the intended services.

Evaluation of Results

Evaluation forms a logical last point in the project cycle whereby results are compared with intentions. Elsewhere, we have described three types of evaluations: namely, built-in, special, and impact evaluations.[26] The first is a routine activity intended to measure a project's actual versus intended progress. Rates of expenditures and staffing are often measures by which management decides if progress is satisfactory.

Special evaluations "are conducted for non-routine reasons, as when some part of the project needs intensive investigation [as when] (1) management encounters problems it cannot solve by itself, (2) an opportunity arises that suggests possible changes in scope or intensity of activities, or (3) something is sufficiently interesting to warrant attention[27]." Impact evaluations are attempts to learn what the project has accomplished. To be useful, enough time needs to elapse for the project's outputs to be felt and measured. Major drawbacks to such evaluations include the time required for the project to be implemented, changing conditions that may invalidate meaningful measurements of accomplishments relative to what conditions would have been without the project, and a lack of interest on the part of host-country officials. USAID has evaluated some of its activities,[28] but drawing firm conclusions from such reviews is not always easy.

Baseline surveys may serve as a basis for evaluating project results, but these can both take up considerable time and divert the project team's efforts. One of the most effective alternative means for evaluating projects that I have come across is FSR's use of farm records. These records, kept to learn about farmers' production activities, have been of considerable value to Guatemala's Agricultural Science and Technology Institute in measuring the acceptability of alternative technologies.[29] Because the Institute works with relatively homogeneous blocks of farmers, a technology's impact on farmers in the area can be estimated from the sample of farmers keeping records. Advantages of this approach are that the data needed for monitoring the on-farm experiments also serve for measuring project impact, experimentation is not delayed until an initial survey is completed, and data requirements are minimized. Moreover, data from control groups not participating in the on-farm experiments can be obtained along with data obtained from farmers participating in the experiments.

PROJECTS AS A MEANS FOR TRANSFERRING TECHNOLOGY

Projects serve the dual purpose of increasing income and welfare through production of goods and services and of providing an opportunity for transferring technology to the project area. For example, a USAID-funded irrigation project for small-scale farmers in Peru concomitantly expanded the area's irrigation network and introduced improved irrigation techniques by funding technical assistance in agricultural research and extension. Such technical assistance contracts commonly accompany USAID grants and loans. Thus, when discussing the linkages between development projects and the central government, we should also be thinking about the opportunities these projects provide for introducing improved technologies to the recipient countries. By improved technology, I refer to design, construction, and operational methods that are more effective in meeting a country's objectives than the technologies currently in use. In the Peruvian case, the improved technology included more effective means of on-farm water management.

The methodologies of FSR offer another means for improving a country's food production. But rather than improving technologies through transfer, FSR offers a means for generating improved technologies within the country.[30] I prefer the FSR approach over the more traditional approach of taking a technology developed in one environment and adapting it to another environment, for two reasons. First, the FSR approach tends to yield a wider range of ideas than otherwise and, second, the combination of disciplines participating in the process is likely to be greater and more integrated.

Concerning the first of these reasons, the on-farm approach to the development of improved farming practices espoused by those such as Byerlee[31], Hildebrand,[32] and Zandstra et al.[33] brings farmers, researchers, and extension staff together in the farmers' environment to identify farmers' problems. There, synergism occurs in which a better understanding of farmers' problems and opportunities for solution are possible than from alternative approaches. For instance, when farmers directly, or through their extension agents, bring problems to the researchers, the farmers' real problems may be incorrectly diagnosed. Bringing the researchers to the field to observe directly the farmers' problems may still not identify the farmers' most pressing problems if the researchers

58

fail to listen to the farmers and extension staff. Alternatively, by allowing the researchers to develop an improved technology (e.g., a shorter-season variety) and then to test it out on the farmers' fields may benefit the farmers, but not as much as had the researchers collaborated with the farmers and extension staff at the offset when deciding on the research agenda.

FSR as currently practiced in many parts of the world allows for such generation of improved technologies. As an example, CIMMYT working with the National Institute for Agricultural Research in Ecuador developed an improved bean-corn combination that met farmers' needs for a shorter-season maize variety with a stalk strong enough to support the climbing beans that were intercropped with the maize. Earlier research had produced corn stalk that could not support the beans' vines.

The other aspect of FSR leading to improved technology generation is the practice of combining disciplines as part of the research process. Those engaged in such research purposefully combine biological, physical, and social scientists in their attempt to identify farmers' problems, in considering alternative solutions, in designing and conducting experiments and studies, and in evaluating and implementing the results.[34] By combining the relevant disciplines in an interdisciplinary manner, more technologies relevant to farmers' needs are bound to be generated than when the disciplines work independently. The synergistic effect is similar to that occurring when researchers, farmers, and extension staff meet in the farmers' environment.

These two FSR concepts--understanding farmers' conditions and interdisciplinary teamwork--have their counterpart in project planning. First, expatriate advisers working with host country staff need to understand the country's conditions, opportunities, and objectives. Then, working together they should be able to come up with appropriate solutions to the country's problems and opportunities. Second, the multifaceted nature of development projects calls for interdisciplinary teamwork. Otherwise, a project may fail because some factor obvious to one of the disciplines was overlooked. Leaving out key disciplines might lead to project deficiencies such as not understanding the motivations of the private sector, misreading the political situation, incorrectly estimating the market for the project's output, or not adequately considering a technical factor such as salinity or sedimentation.

ISSUES

This chapter closes with four questions about development projects followed by some thoughts on possible answers. These questions are not exhaustive and the answers are more to stimulate thought than anything else.

Some Problems with Development Projects

1. How do planners in a developing country identify relevant projects and avoid unduly wasting time and effort on poor project possibilities?
2. How can these planners handle the complexity so often encountered with development projects?
3. How can those responsible for designing development projects manage interdisciplinary teams.
4. How can the central government make a significant impact on the country through the use of development projects?

Some Possible Answers

Beginning with the first question about wasting time with poor projects, I have argued in this paper that development projects are an integral part of a country's macroeconomic planning process. Therefore, the projects should evolve from a consideration of the country's problems, opportunities, and objectives. At the same time, development projects must be viable technically so that they function reasonably well biologically and physically, produce an output that is in demand, and meet the economic, financial, and other requirements of the participants. Also, by maintaining a modicum of caution, planners should be able to guard against misrepresentation by potential investors. When planners pay attention to these types of factors, they should be able to screen out many of the unattractive projects before spending too much time on them.

Another way to reduce the risk of project failures is to introduce projects on a limited scale. This approach is particularly adaptable when a project comprises a set of replicable activities. When introduced sequentially, the more limited program will be more manageable, project staff can be trained as the program evolves, and the experience gained can be used when expanding the project to its full size.

For example, the small-scale irrigation project in Peru was intended eventually to cover the principal irrigation areas throughout the Andean valleys. Rather than attempting to implement a country-wide program, we proposed two project areas. The first was to be close to the agency's headquarters, where access would be easy and chances of success were seemingly the highest. This project would be followed six months later by a second one, where conditions were somewhat different.

Similarly, a project design team from Colorado State University proposed new FSR activities in two regions of Tanzania before attempting to implement the FSR program in each of the country's six major agroecological regions. While such stepwise expansion may appear perfectly logical to the planner, a country's leaders frequently press for nationwide programs in

60

their desire to show major accomplishments quickly and to guard against claims of favoritism to particular regions or groups.

Concerning the second question, planners need to recognize the complexities of projects by allowing enough time for identification, design, analysis, and implementation. Interdisciplinary teams provide a better means for understanding project complexities; and the practicalities of meeting deadlines and cost constraints often dictate that project planners search for improvements that are not necessarily optimal. Such a pragmatic approach reduces the time and complexity of activities, focuses attention on a project's most important facets, and is often responsive to the capabilities of the country's planners.

Project complexity can also be reduced by minimizing the amount of detailed planning whenever decisions about particular activities can be delayed without seriously jeopardizing the project's outcome. This approach is a frequently practiced management tool and conforms to Hirschman's "principles of the hiding hand."[35] Hirschman argues that detailed planning is never completely satisfactory, since so many variables influence the project's outcome. Consequently, planners are better off concentrating on a limited number of key factors and leaving decisions on the rest until later. By waiting to make such decisions, the requirements of future decisions will be better understood and the ideas for solutions will generally be more appropriate.

Hirschman's "principle of the hiding hand" includes the concept that planners tend to underestimate the "various difficulties that lie across the project's path, on the one hand, and of the ability to solve these difficulties on the other hand." Moreover, excessive planning for future events produces the illusion that the planners have carefully thought-out solutions, whereas such planning often obscures the planners' failure to clearly focus attention on the key factors for a project's success.

Next, managing interdisciplinary teams is not an easy task. Those of a discipline have their own paradigms, which in turn influence attitudes--some of which do not facilitate interdisciplinary teamwork. Understanding and respecting another's disciplines takes time and requires motivation. Also, finding ways to adequately reward team effort has its problems. Nevertheless, interdisciplinary teamwork is important to project planning and needs to be promoted. This can be accomplished by (1) selecting team members who have an inclination for teamwork, (2) providing team members the chance to interact in identifying problems and opportunities and in proposing and testing solutions, (3) urging team members to learn and appreciate the paradigms of other team members, and (4) adequately recognizing and rewarding team accomplishments.

A technique employed by FSR teams in Guatemala and Honduras is for teams to identify the overriding problems of farmers representative of a specific type of farming system and

61

to measure the team's output by the extent to which the team's proposed improvements are adopted by the farmers. Tendencies for specialists to polarize according to their own disciplines is guarded against by (1) having the team initially focus its attention on the whole farming system and (2) encouraging team members to identify problems and offer solutions outside their own discipline. A product of this approach is that soon an outside observer may find it difficult to identify a team member's discipline. Agronomists may speak of marginal returns to labor and agricultural economists may speak of split-plot designs--and both will know what they are talking about.

Finally an Achilles' heel of project work is the development of schemes that concentrate too much of a country's scarce resources on projects yielding too few results. In my work with the planning commission in Ethiopia, we sought replicability of results; and in FSR, "homogeneous" farmers are identified so that successful experiments with the participating farmers can be applied to other farmers with similar characteristics.

Another way to promote a country's growth through development projects is by expanding the level of project participation in which the central and regional governments work with private, national, and international groups. Too often, private and international funds go unused because a country cannot generate enough projects that meet the funding organizations' standards. This bottleneck refers to a country's absorptive capacity, which was so commonly discussed a number of years ago. Imaginative use of the resources of these other groups will go far to increase the size of a country's development program. For this to occur those in central planning will need to bring this opportunity to the attention of their country's top decision makers.

REFERENCES

[1]Warren C. Baum, "The Project Cycle." Finance and Development 7 (1970):2-13.

[2]John A. King, Economic Development Projects and Their Appraisal (Baltimore, Md: The Johns Hopkins University Press, 1976).

[3]Albert Waterston, "A Hard Look at Development Planning," Finance and Development 3 (1966):85-91.

[4]Baum, "The Project Cycle."

[5]Lester R. Brown, By Bread Alone (New York: Praeger Publishers, 1974).

[6]Sterling Wortman and Ralph W. Cummings, To Feed this World: The Challenge and the Strategy. (Baltimore, Md.: The Johns Hopkins University Press, 1978).

[7]H. E. Gilbert, D.W. Norman, and F. E. Winch, Farming Systems Research: A Critical Appraisal, MSU Rural Development Paper no. 6 (East Lansing, Mich.: Department of Agricultural Economics, Michigan State University, 1980).

[8]Paul N. Rosenstein-Rodan, "Notes on the Theory of the Big Push, in Readings in Economic Development, eds. Theodore Morgan et al. (Belmont, Calif.: Wadsworth, 1963).

[9]Albert O. Hirschman, The Strategy of Economic Development. (New Haven, Conn: Yale University Press, 1958).

[10]W. Arthur Lewis, Development Planning. (New York: Harper and Rowe, 1966).

[11]W. Arthur Lewis, Report on Industrialization and the Gold Coast (Accra, Gold Coast: Government Printing Office, 1953).

[12]Peter G. Sassone and William A. Schaffer, Cost-Benefit Analysis: A Handbook (New York: Academic Press, 1978).

[13]W.W. Shaner, Project Planning for Developing Economics (New York: Praeger Publishers, 1979).

[14]Donald Winkelmann and Edgardo Moscardi, "Aiming Agricultural Research at the Needs of Farmers," in Readings in Farm Systems Research and Development, eds. W. W. Shaner et al. (Boulder, Colorado: Westview Press, 1982)

[15]A.C. Zulberti, K.G. Swanberg, and H.G. Zandstra, "Technology Adaptation in a Colombian Rural Development Project," in Economics and the Design of Small Farm Technology, eds. A.G. Valdez, et al. (Ames, Iowa: Iowa State University Press, 1979).

[16]Julius Margolis, "Secondary Benefits, External Economies, and the Justification of Public Investment," Review of Economics and Statistics 39 (1957):284-291.

[17]Michael Roemer and Joseph J. Stern, The Appraisal of Development Projects: A Practical Guide to Project Analysis with Case Studies and Solutions (New York: Praeger Publishers, 1975).

[18]Lan Little and James A. Mirrlees, Project Appraisal and Planning For Developing Countries (London: Heinemann Education Books, 1974).

[19]Lyn Squire and Herman G. van der Tak, Economic Analysis of Projects (Baltimore: The Johns Hopkins University Press, 1975).

[20]United Nations Industrial Development Organization, Guidelines for Project Evaluation (New York: United Nations, 1972).

[21]Colin Bruce, Social Cost-Benefit Analysis: A Guide for Country and Project Economists to the Derivation and Application of Economic and Social Accounting Prices, World Bank Staff Working Paper no. 239 (Washington, D.C.: International Bank for Reconstruction and Development, 1976).

[22]Deepak Lal, Wells and Welfare: An Exploratory Cost-Benefit Study of the Economics of Small-Scale Irrigation in Maharshtra, Development Centre of the Organization for Economic Cooperation and Development, Series on Cost-Benefit Analysis, Case Study, no. 1 (Paris, 1972).

[23]Ian Little and D.G. Tipping, A Social Cost-Benefit Analysis of the Kola Oil Palm Estate: West Malaysia, Development Centre of the Organization for Economic Cooperation and Development, Series on Cost-Benefit Analysis, Case Study, no. 3 (Paris, 1972).

[24]J. Price Gittinger, Economic Analysis of Agricultural Projects (Washington, D.C.: International Bank for Reconstruction and Development, 1981).

[25]Instituto de Ciencia y Tecnologia Agricolas, Objetivos, Organisacion, Funcionamiento, Publicaciones Miscelaneous, Folleto no. 3 (Guatemala City, 1976).

[26]W.W. Shaner, P.F. Philipp, and W.R. Schmehl, Farming Systems Research and Development: Guidelines for Developing Countries (Boulder, Colorado: Westview Press, 1982).

[27]Ibid.

[28]United States Agency for International Development, Columbia: Small Farmer Market Access, USAID Project Impact Evaluation Report no. 1 (Washington, D.C., 1979); Kitale Maize: The Limits of Success, USAID Project Impact Evaluation Report no. 2 (Washington, D.C., 1980); Impact of Rural Roads in Liberia, USAID Project Impact Evaluation Report no. 6 (Washington, D.C., 1980).

[29]Peter E. Hildebrand , The ICTA Farm Record Project with Small Farmers; Four Years Experience (Guatemala City: Instituto de Ciencia y Tecnologia Agricolas, 1979).

[30]H.E. Gilbert, D.W. Norman, and F.E. Winch, Farming Systems Research: A Critical Appraisal MSU Rural Development Paper no. 6 (East Lansing, Mich: Department of Agricultural Economics, Michigan State University, 1980).

[31]D. Byerlee, M.P. Collinson, R.K. Perrin, D.L. Winkelmann, S. Biggs, E.R. Moscardi, J.C. Martinez, L. Harrington, and A. Benjamin, Planning Technologies Appropriate to Farmers: Concepts and Procedures (El Batan, Mexico: CIMMYT, 1980).

[32]Peter E. Hildebrand, Generating Small Farm Technology: An Integrated Multidisciplinary System (A paper presented to the 12th West Indian Agricultural Economics Conference, Caribbean Agro-Economic Society, Antigua, April, 1977), pp. 24-30.

[33]H. G. Zandstra, E. C. Price, J. A. Litsinger, and R. A. Morris, A Methodology for On-Farm Cropping Systems Research. (Los Banos, Philippines: IRRI, 1981).

[34]W. W. Shaner, P. F. Philipp, and W. R. Schmehl, Farming Systems Research and Development: Guidelines for Developing Countries. (Boulder, Colorado: Westview Press, 1982.

[35]Albert O. Hirschman, Development Projects Observed. (Washington, D.C.: The Brookings Institution, 1967).

6

Inducing Development at the Micro-level: Theory and Implications for Technology Transfer Strategies

Robert D. Stevens

INTRODUCTION

This chapter examines the process of technology transfer by first highlighting the very different economic, technical and institutional conditions found in developing countries. A review of the economic theory of induced technical change is then undertaken for insights into effective strategies for technology transfer today. The implications of the induced technical change model for strategies in phases of technology transfer are brought together in a final section.

As a preliminary matter, there is little disagreement that the objective of technology transfer is to aid in accelerating growth in production and income. It is also understood that double, four-fold and eventually ten-fold increases in farm labor productivity can be achieved through effective technology transfer.

Most also recognize that agricultural technology change is generally a blunt instrument for greatly affecting the relative distribution of income, as Schuh has emphasized recently.[1] However, technology transfer is likely to cause changes in the distribution of income. Where new technology is adopted, some will gain and some will lose, at least relatively. If a capital intensive technology is adopted in a farming area, those who supply that technology to farmers are likely to gain, and laborers are likely to lose. If it is a labor intensive technology, laborers will probably gain relative to others.

Most are aware that in a large number of developing nations income distribution is much more highly skewed than in many of the more developed nations. Yotopoulos and Nugent provide data indicating that in many higher income nations, the top 20 per cent of persons receive 40 to 50 percent of national income while in lower income nations, this group of persons receives 50 to 70 percent of national income.[2] We therefore often are concerned that the new technology contribute to a reduction in the skewness of the income distribution, if possible.

The desires and values of the people who are the recipients of the technology should have the most influence on decisions about adoption of new technologies. How would they view the likely income distribution effects of the new technology? We should at least be able to identify groups that would probably benefit and those that might lose. I believe technology transfer without some estimate of its income distribution implications is no longer professionally responsible. Hence, examination of actions which could be used to assist adjustment and to compensate losing groups may also be called for.

A second preliminary point is that the assessment of the potential impact of new agricultural technology in rural areas of developing nations needs to be undertaken in a broad framework which includes the longer-term national economic development goals which are often incorporated in development plans. Thus, the impacts of technology transfer need to be related to the longer-term dynamics of development. Hence, such variables as rates of urbanization, regional rural population growth, industrialization, and general rural development goals need to be kept in view as ex ante technology assessment is undertaken.

There are two general hypotheses addressed in this chapter. First, that at this stage in the transformation of agriculture in developing nations, the direct transfer of agricultural technology in the form of materials currently used in more developed nations will provide limited increases in food production. Second, therefore, adaptive design transfer activities carried out largely in developing nations will contribute significantly to greater food production in the next two decades.

CONDITIONS AFFECTING TECHNOLOGY TRANSFER

Variability Among Developing Nations

Perhaps the most important perspective that needs emphasis in a discussion of the transfer of food production technology is the wide variability in economic and technical conditions, our relative ignorance about institutional conditions in developing nations and how little we still know about farm production conditions in areas where small farmers struggle to attain a livelihood. Detailed knowledge of particular environments is essential for identification of profitable opportunities which will induce technical transfer and development.

Most agricultural areas of developing nations are still in the beginning phases of a century-long agricultural and rural transformation. These highly variable low-income areas have one characteristic in common: their economic, technical, and institutional environments are very different from those in the more developed nations.

Economic differences in many of these countries that contrast greatly with conditions in more developed nations include particularly high capital costs relative to labor costs, especially high machinery costs. In more densely populated nations, land costs are usually much higher relative to labor costs, as compared with the United States for example.

Two other very important differences impinging on technology transfer in less developed nations are the small size of farms and the limited number of domestic firms which have the ability to develop and produce mechanical or biological technology for agriculture. In addition, the supply of individuals who have technical and scientific knowledge in these countries is small and largely inelastic in the short run. Important implications about effective technology transfer which follow from these economic differences are explored in detail below.

The differences from more developed nations in physical and biological environments are generally well recognized, such as tropical versus temperate or subtropical climates and vastly different pest and weed conditions. Also, widely different institutions and social agreements, which determine the way things are done, place other constraints on attempts to introduce new, more productive technology.

The recognition of these vastly different environments points to the importance of gathering detailed information about the economic, technical, and institutional realities of the target areas for effective technology transfer. Large amounts of farming systems research, or similarly detailed studies of the economic and technological environment, and of the institutional and cultural factors which directly affect agricultural production, are therefore required to enable effective technology transfer to be carried out. Some of the more effective and greatly increased activities in this area are currently being led by a number of the International Agricultural Research Centers.[3]

Phases of Technology Transfer

The material, design and capacity transfer phases of technology transfer outlined by Hayami and Ruttan may by now be well known.[4] They provide a useful categorization of transfer activities to aid in identifying appropriate strategies for transfer of technology into particular less-developed agricultural areas, and are summarized here.

Material transfer. This phase involves transferring from more developed countries such materials as seeds, plants, machinery, pesticides, and fertilizers. If the material technology which is transferred cannot demonstrate a fairly high return, perhaps 20 to 30 percent on investment, farmers are not likely to become interested because of the costs of change and the risks and uncertainties associated with incorporating new production practices into farming.

We hypothesize that today, after 30 years of accelerated world agricultural development, in those developing nations where relatively free trade has been present, in a large proportion of the agricultural areas of these nations, there remain relatively few opportunities where new agricultural

69

materials imported from developing countries would be very profitable. This hypothesis implies that importers and entrepreneurs have been carrying out material transfer activities for considerable periods of time and continue to be actively engaged in material transfer activities.

Design Transfer. In this phase the designs, blueprints, formulas, and books used by scientists and technologists are transferred to the target country environment. This activity can involve book translations, english language training for scientists and technologist and short-term training abroad up to and including masters-level training. Through collaborative orderly testing, multiplication of seed, adaptive research and particularly tests on farmers' fields, modifications in technology obtained from more developed nations are made, resulting in many profitable new investment opportunities for farmers in the target environment.

This phase includes domestic multiplication of plant materials, expanded local production of machinery and parts, and building and strengthening experiment stations. We hypothesize that a very large number of opportunities for high returns to design transfer activities are present in many less developed nations today.

Capacity Transfer. Capacity is the most significant investment activity in technology transfer. In addition to all the activities in the previous phases, it requires the production of significant numbers of Ph.D's, or their equivalents, who can provide scientific and technical leadership in the private and public sectors for national technology development. Attractive professional working environments have to be created which include large libraries, laboratories, computers, and other equipment to attract and retain highly qualified people and assure their productivity. Salaries sufficiently high to draw and retain top talent with sufficiently free working environments are needed so these professionals can concentrate on scientific matters. When capacity transfer has been achieved, professionals in developing nations are able to create crop varieties for local conditions and new agricultural machines which prove attractive to farmers. They also contribute at the scientific frontier.

The questions of priority, or relative allocation of resources to each phase of technology transfer at a particular time, in a specific country or region, for identified crop, livestock or fish enterprise(s) depends upon the amount of activity which has already been undertaken in the three phases. We hypothesize that in most countries in the last thirty years, large amounts of material transfer have already occurred because material transfer is the easiest. Thus, we would expect that the design phase is where high returns are likely to be obtained in technology transfer at this time.

Due to the long term pay-off from capacity transfer, some resources should be devoted now to planning a program of planning a program of capacity transfer in a long-time horizon of 20 to 30 years. In order to productively use national resources, the plans for capacity transfer need to be realistic for the particular developing nation given its size, the expected importance of different agricultural enterprises, and the international research environment with which the nation can interact.

INDUCED TECHNICAL CHANGE

To identify the reasons for our hypotheses about technology transfer, we briefly review the economic theory of traditional agriculture and of the economic inducement mechanism for technology development.

Farmers' Responsiveness to Profitable Technology
T. W. Schultz's theory of economic equilibrium in traditional agriculture may, by now, be familiar to all.

> The critical conditions underlying this type of equilibrium, either historically, or in the future, are as follows: (1) the state of the arts remains constant, (2) the state of preference and motives for holding or acquiring sources of income remains constant, and (3) both of these states remain constant long enough...to arrive at an equilibrium...[5]

Although many changes in rural areas of developing nations may be progressively upsetting this equilibrium, our concern is with the extent to which "the state of the arts," or technology, has changed. The theory of an economic equilibrium is dependent upon the hypothesis that traditional farmers are economically rational, given their personal talents and health, their resources and the institutional and cultural constraints within which they operate. Much research on agriculture in developing nations in the last 15 years has demonstrated the economic rationality of these often illiterate farmers, and that they will respond positively to demonstrated opportunities to obtain increased net income, with appropriate discounts for risk and uncertainty.[6]

Thus, in examining the agricultural technology used by farmers in developing nations today, when we find that farmers continue to use traditional agricultural techniques, or some which appear to be overly labor intensive, our hypothesis must be that these technologies are the most profitable available to them and that the cost and/or risk of changing to new, potentially more profitable technologies, must be too great. The challenge presented to those intent upon technology transfer is, therefore, to demonstrate on representative farms, significantly more profitable agricultural technology than that in current use.

71

Why Material Transfer is Often Unprofitable

Relative factor prices and the Hayami and Ruttan model of induced technical change show why so much material transfer of agricultural technology from more advanced nations is not profitable on small low-income farms in developing nations.[7] To demonstrate this, we will review the economic theory of induced technology development, particularly as presented by Evenson and Binswanger.[8] The essence of the argument is that in many developing nations, low cost labor and high cost capital make the highly capital intensive agricultural technologies of more developed nations unprofitable in less developed nations.

The different ratios between the cost of varied resources used in agricultural areas around the world cause varied technologies to be profitable in diverse locations. Different technologies are induced to be developed by dissimilar relative prices in the various economic environments. Specifically, in the large number of low income nations with labor costs low relative to capital costs, labor-intensive technologies are more profitable than capital-intensive technologies. These realities can be demonstrated using production isoquants (Figure 6.1).

A traditional, or low productivity technology is shown as isoquant a, with an equilibrium point L, with a certain ratio of the cost of labor to capital indicated by isocost line E. In this economic environment improved technologies b and c would be induced to be imported or developed. They would provide the same output with less resource cost, and hence greater profit. New technologies of this sort for a major crop, if patentable, would provide large profits to the innovator if in the private sector. Increased prestige and probably larger budgets to the research station or workshop would likely follow if the innovator were in the public sector.

With the same relative prices for labor and capital as previously, the parallel price line F indicates that equilibrium points M and N would be indifferent economically as to which technology was used. If considerable unemployed labor were available, farmers might choose the more labor intensive technology to aid in supporting their relatives and fellow villagers.

In the United States and other high labor cost nations, the technologies which have been induced to be developed have led to increasing substitution of relatively lower cost capital for labor. It is of some interest that in both the United States and Japan, the cost of today's labor relative to capital as represented by agricultural machinery has increased about three times its 1880 level. (Table 6.1).

This change in the ratio of labor to capital costs indicates that the price line for nations with high cost labor becomes steeper (Line G in Figure 6.1). This relative price line was drawn by increasing the cost of labor three times relative to capital, so that only one third the labor can be

Figure 6.1. Induced Technical Innovation with Low and High Labor Costs
Relative to Capital

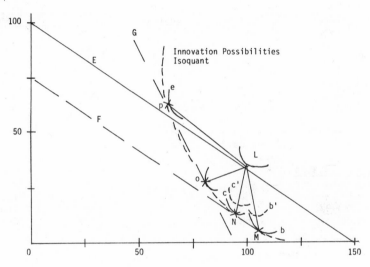

Cost of Labor

TABLE 6.1

Change in the Cost of Labor Relative to Capital as Represented
by Agricultural Machinery in Japan and the United States, 1880
to 1960

Country and Year	Labor Cost (Farm Wage)	Machine Cost	Ratio of Labor Cost to Machine Cost	
			Ratio	Index
	(Yen per Day)	(Yen)	(1000)	
Japan				
1880	.22	66	3.3	100
1920	1.39	160	8.7	263
1960	440.00	37000	11.9	361
	($ per Day)	($)		
United States				
1880	.90	146	6.2	100
(1925)[a]	2.35	152	15.5	250
1960	6.60	356	18.5	298

[a] The year 1925 is more representative of long-run changes in
prices than the year 1920 for the United States.

Source: Adapted from Yujiro Hayami and Vernon W. Ruttan, Agricultural Development: An International Perspective
(Baltimore: The Johns Hopkins University Press,
1971), Table C-2, p. 338.

obtained for a given amount of money as compared with price line E. The changed labor to capital cost ratio indicated by Line G would induce the development of technologies represented by isoquants d and e, which employ increased amounts of capital and reduce labor, with equilibrium points O and P. If a curve is drawn through the equilibrium points of the isoquants g, c, d, and e, an innovation possibility isoquant for all possible price ratios of capital and labor can be drawn. It should be emphasized that beforehand it is difficult to estimate how long the arrows will be and what labor-capital ratio individual technologies will require. The skill of scientists, their resources and the nature of technological advance in given time periods will determine this.

Technology Transfer Under Low Labor Cost Conditions.

This analytic framework shows why much of the present technology in developed nations is not likely to be profitable in low labor cost developing countries. Today's technologies in the more developed nations are likely to be represented by isoquants d and e. As these isoquants are above the price line F, which indicates the relative price of labor to capital in low labor cost developing nations, they would be less profitable to use in developing nations than technologies b and c.

A point of particular interest related to material transfer arises here. Have technologies b and c been developed for a particular crop on agricultural area? If technologies b and c are not available and technology d could be transferred from a more developed nation, its adoption would reduce resource costs and increase income. This would be an example of the material phase of technology transfer, which we hypothesize has been occurring in a fairly widespread manner in the last thirty years. Technology d, however, is likely to have serious drawbacks. It decreases employment as compared to technology a. Despite this, the transfer of some more developed country technologies represented by technology d has and will contribute to increased agricultural growth in low income nations.

The greater challenge in technology transfer is to develop, through design phase activities, the more profitable lower cost agricultural technologies represented by isoquants b and c for the agricultural areas of the large number of low labor cost nations. Labor costs in many of these nations are expected to remain low for a generation or two.

In passing it should be noted that this analysis has been limited to considering different relative prices for labor and a generalized measure of capital. Similar analyses might be carried out usefully, relating land costs to labor costs, and the costs of particular capital items such as different agricultural machines to labor costs, and also the cost of other resources such as mechanically produced irrigation water to labor costs. In this way ex ante analysis could examine the resource use implications of material and design phase technol-

75

ogy transfers into particular farming regions in developing nations.

The Role of Design Transfer.
Pursuing the analysis further, technology development aimed at reaching isoquants b and c employing design transfer methods has certain limitations and hence technologies b^1 and c^1 may only be reached by these methods. These technologies may not be very much more profitable than technology d borrowed from a more developed nation as they may reach approximately the same isocost line, but they would be preferable as capital saving and labor using technologies, thereby providing more income to poorer segments of the rural population.

The achievement of the capacity level of technology innovation in country or through arrangements with other countries or the international agricultural research institutes will enable less developed nations to obtain technologies b and c, the most profitable possible, given the existing level of agricultural science available in the world at any given time and local relative prices for inputs.

It is of interest to note also that the modern technology e, which is a least cost technology in high labor cost nations, would actually increase costs and reduce the income of traditional farmers if they were persuaded to shift from technology a. How often have aid programs attempted to transfer this kind of capital intensive modern agricultural technology to farmers in less developed nations where labor costs are so very much lower?

From this review of the induced technical innovation mechanism we can see that much of the agricultural technology developed in the last thirty years in high labor cost nations, is likely to be represented by isoquants d and e, while the most profitable technologies for low labor cost nations are represented by isoquants b and c. This conclusion leads to the question of how, in the shorter run, technology transfer activities can aid in developing these most profitable, labor intensive technologies.

The analysis has focused on nations with low labor costs. It should, however, be recognized that in some less developed nations, the price ratios for capital and labor are closer to those in more advanced nations. In these circumstances, capital intensive technologies d and e are more likely to be profitable. Nations with these conditions may include a number of the oil exporting nations. In these nations, agronomic and other technical and institutional issues may be more crucial to the profitability of technology which may be transferred.

Paths of Technology Development
Perspective on the very different paths of technology development which have been induced to be followed by different nations was provided by Hayami and Ruttan using labor and land

productivity for different nations (Figure 6.2). Relatively labor-short, land-rich nations, such as the United States adopted agricultural technology which greatly increased labor productivity, while only slightly increasing land productivity. In contrast, land short nations, such as Taiwan and the United Arab Republic, have used technologies which increased land productivity, without increasing labor productivity very much. To take specific examples, in 1880, the United States is estimated to have produced about 15 wheat units per male worker in agriculture (a wheat unit is the equivalent of one ton of wheat). At this time, the United States achieved a land productivity of only about .5 wheat units per hectare. In contrast, at the same time, Japan is estimated to have produced an annual agricultural output of 2.4 wheat units per male worker and 2.7 wheat units per hectare of land. Thus in 1880, labor productivity in agriculture in the United States was more than five times as great as in Japan, while in Japan land productivity was almost five times as great as in the United States.

By 1960, the United States had increased its labor productivity to almost 100 wheat units per male worker in agriculture. While land productivity had only increased a little over the period to .8 wheat units per hectare. During the same period Japan increased its labor productivity to nearly 11 wheat units and its land productivity tripled, to 7.5 wheat units per hectare. The land and labor productivities of other selected nations are also shown. The intermediate positions are represented by a wide range of nations including most European countries which have employed agricultural technologies which have increased both land and labor productivity considerably. In the next 80 years we can expect many less developed nations to achieve similar large increases in agricultural productivity, along technology paths induced by the relative cost of capital to labor in their economics. This will occur as technology transfer moves through the material, design and capacity phases in each less developed nation.

A final point drawn from this theory is that the technologies of many of the more developed nations which have followed intermediate paths in technology use may be more profitable in labor surplus, land short developing nations than U.S. agricultural technologies. This suggests that we may need to put more effort into securing a wider international framework for technology transfer efforts. Confirming attention to the transfer of U.S. farming technology, which is less likely to be profitable in many developing countries could reduce the cost-effectiveness of technology transfer activities.

STRATEGIES IN THE THREE PHASES OF TECHNOLOGY TRANSFER

The implications of the theory of induced technical innovation for strategies of technology transfer will be considered under four headings: material, design, and capacity transfer

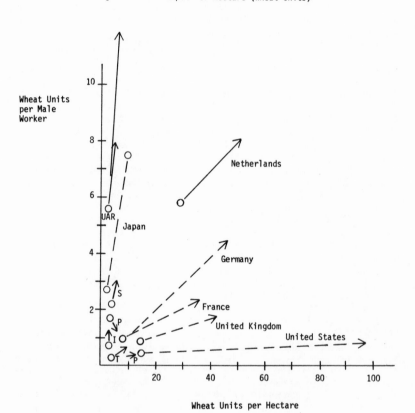

Figure 6.2. Paths of Growth in Land and Labor Productivity
for Selected Countries, 1880 - 1965

Agricultural Output Per Hectare (Wheat Units)

Wheat Units
per Male
Worker

Netherlands

UAR
Japan

Germany

S

France

P
United Kingdom

I
United States

T
P

Wheat Units per Hectare

and wider implications. Our focus is upon the requirements for effective technology transfer in each phase. Comments will be limited to economic and institutional issues, leaving aside the large number of technical issues.

Material Transfer

For developing nations where import and communication barriers have been low, we expect at this time that large amounts of the more profitable material transfers have already taken place or are proceeding involving a considerable number of international and domestic firms. These activities are due to economic incentives for international transfer of technology and the fact that material transfer is the least costly of the three levels of technology transfer.[9] This working hypothesis leads to high priority being placed on two activities. One activity is identification of the amounts and kinds of material transfer activity currently taking place to determine the extent of material technology flows, the supply of firms engaged in material transfer and to determine if the personnel supply for this activity is fairly elastic as hypothesized.

The second activity is farm level research. Effective work at all three levels of technology transfer requires detailed knowledge of representative farms through surveys and comparative testing. Objectives of farm level research include identification of: 1) the technical and economic characteristics of the technologies currently being used; 2) the most pressing technology needs; 3) estimates of the costs and supply elasticities of resources (inputs) used by farmers--capital, labor (particularly at peak labor demand periods), and land; 4) significant price distortions, so that efforts may be directed toward institutional change and/or change in national price policies, if deemed feasible and desirable.

Comparative tests of the performance of currently used technology transferred from other agricultures is also essential for decisions about design transfer activities. Although tests on experiment stations are useful for some purposes, the performance and cost estimates obtained under usual farm operating conditions on representative farms is a core requirement for effective technology transfers. Any new technology will in most cases need to be appreciably more profitable, providing perhaps a 15 to 25 percent increase in return to compensate for the costs of learning and change. Technologies which are less costly for the farmer to adopt will require a smaller increase in profit to induce use.

Very important information from comparative tests include estimates of the shifts in the ratios of resources required by the new as compared to currently used technology. Knowledge of expected significant changes in capital requirements or labor use, and how labor use patterns are shifted in relation to peak labor demand periods is needed to facilitating technology change. The recent surge in farming systems research, lead partly by the international agricultural research institutes, provides useful methodology for these activities.[10]

Design Transfer

The microeconomic theory reviewed and the experience of the last thirty years suggests that further effort in the design phase of technology transfer is likely to be highly productive at this time in many developing nations. A large number of technologies which would be profitable in low labor cost developing areas have probably not been developed due to a low level of domestic design capacity both in experiment stations and in agricultural input industries.

Achieving design transfer abilities often requires intermediate run cooperative agreements and working arrangements with public and private domestic and overseas organizations[11]. For design transfer in biological technology, cooperation with national and regional experiment stations will be the usual route, although there may be private seed companies, or fertilizer and pesticide firms that could rapidly develop and test new technology on farmers' fields. For design transfer in mechanical technology a variety of public and private institutions will be most usefully worked with, from nascent agricultural machinery suppliers to a wide range of public and quasi-public units including experiment stations.[12] Some technical problems, if clearly identified on the basis of less developed nation experience, can be productively worked on in U.S. agricultural experiment stations, but the bulk of high-return adaptive and developmental research in the design transfer phase is likely to be much more effective if carried out in developing nations.

Human capital investment at this stage of technology transfer may involve short-term technical training overseas and considerable in-country training to increase abilities to use the designs, books and other design level materials which have been transferred. Some M.S. level training will be likely to contribute to effective design level technology transfer. Successful design transfer necessarily includes large amounts of representative on-farm research and testing is outlined above under material transfer.

Capacity Transfer

Capacity transfer is the longer range high pay-off objective of technology transfer. To be cost-effective, this objective has to be carried out in relation to the evolving international network of agricultural technology creating institutions and firms, in relation to national resources available, and the national importance of the different agricultural enterprises. Because of economies of scale in much capacity-level research and technology development, many smaller nations will have resources sufficient only for the mounting of capacity level research related to the most important agricultural enterprises.

At this time, planning and some resource use for further capacity level development needs to be carefully considered. For if the long run pay-off is as high as research has indicated, this long term high-return investment should be commenced. The focus of activities would be two-fold, to train appropriate numbers of Ph.D's, or their equivalents, and secondly, to develop a research environment where scientific frontiers can be reached. Such an environment in more developed nations has included considerable investment for libraries and laboratories, but more important, the stable support of substantial numbers of trained scientists at attractive salaries and the development of an administrative structure which is able to provide direction and incentives for useful research. Graduate training at the advanced level aids development of capacity level technology. Government policies and financial support for success at this level of technology transfer needs to be long run and relatively stable.

Wider Issues

Effective technology transfer, I believe, will require addressing wider issues which impact directly on adoption of more productive technology. Three of the more important wider issues relate to price policies, to institutional change, and to income distribution.

The price policy issues revolve around any significant divergence at the farm level between the social and the private cost of agricultural inputs. So many developing nations have government policies which introduce wide distortions in the prices farmers pay and receive, that in some cases reduction of price distortions may be critical to transfer of more productive technology. For example, high tariffs preventing use of critical technologically advanced agricultural inputs are common.[13]

A part of the Hayami-Ruttan induced development model hypothesis is that institutional change will be induced by technical change.[14] If this hypothesis is valid, successful introduction of new technology will lead to institutional changes. Can these be foreseen and could they be facilitated? What changes in governmental institutions, rules, regulations, etc. would make the new technology more profitable and hence aid in accelerating its adoption? Effective technology transfer is likely to require addressing institutional change issues directly.

I believe socially responsible technology transfer activities require that knowledge be developed beforehand of possible impacts of technology introduction on the different groups in society. With this information, judgments can be made about whether the likely impacts will be significant enough to call

for response by government. Governmental response might include compensation or aid in adjustment. Such policies were
considered by Schultz and included offers of alternative
employment opportunities to agriculture laborers displaced by
technological progress.[15] Additional government resources for
a range of kinds of education, or other activities to aid the
transfer of people to other sectors and areas with better prospects for employment are additional examples.

CONCLUSIONS

This review of technology transfer in the light of microeconomic theory of induced technical innovation has led to the
following five sets of conclusions.

The complex interweaving stream of material technology
transfer will continue to increase in the world, as technology
breakthroughs are achieved in different countries. Productive
opportunities to aid in the further enhancement of current
material transfers to various developing nations may still be
present, but are less likely in nations which have had low
trade barriers. Useful material transfer activities are most
likely in those developing nations which have in the recent
past had high trade barriers which slowed material transfer, or
in the more remote farming areas of developing nations.
Efforts to accelerate material transfer need to focus on aiding
augmentation of self-sustaining private and government networks
which will assure continuous, appropriate rates of material
transfer.

At this time, design transfer activities appear most
likely to contribute to rapid acceleration of agricultural production in many developing nations. Effective design transfer
activities require a high proportion of work to be carried out
in host countries under cooperative agreements with a variety
of organizations including agricultural experiment stations.

As capacity building activities are likely to have high
returns over the longer run, planning for and carrying out some
activity in this phase of technology transfer at this time is
likely to be highly productive. This activity will normally
include selected overseas doctoral-level training.

At all levels of technology transfer, the development of
the capacity for continuing farm level evaluation trials of
technology is essential. The vast differences in economic,
technical and institutional environments of the large numbers
of small farms in the various regions in developing nations
leads to the need for a much larger fund of information about
factor prices and the productivity of food production technology currently used on farms.

Effective and socially responsible technology transfer has
wider, more general requirements which include: 1) obtaining
knowledge of factor and product price distortions for possible
modification, 2) identification of the implications of technol-

ogy adoption for institutional changes and how they might be accelerated, and 3) knowledge of possible significant income distribution impacts and consideration of possible compensatory and adjustment facilitating policies.

REFERENCES

[1]G. Edward Schuh, "Approaches to 'Basic Needs' and to "Equity" Which Distort Incentives in Agriculture" in Distortions of Agricultural Incentives, ed. Theodore W. Schultz (Bloomington: Indiana University Press, 1978), p. 319.

[2]Pan A. Yotopoulos and Jeffrey B. Nugent, Economics of Development: Empirical Investigations (New York: Harper & Row, Publishers, 1976), p. 240.

[3]For example Derek Byerlee et al., Planning Technologies Appropriate to Farmers: Concepts and Procedures (Londres, Mexico: CIMMYT (Centro Internacional de Mejoramiento de Maiz y Trigo), 1980), and M.P. Collinson, Farming Systems Research in Eastern Africa: The Experience of CIMMYT and Some National Research Services, 1976-81, MSU International Development Paper no. 3 (East Lansing, Michigan: Dept. of Agricultural Economics, Michigan State University, 1982).

[4]Yujiro Hayami and Vernon W. Ruttan, Agricultural Development: An International Perspective (Baltimore: The Johns Hopkins Press, 1971), p. 175.

[5]Theodore W. Schultz, Transforming Traditional Agriculture (New Haven: Yale University Press, 1964), p. 29.

[6]A sample from the large literature on allocative efficiency includes: Pan A. Yotopoulos and Jeffrey B. Nugent, Economics of Development: Empirical investigations (New York: Harper Row, 1976), Chapters 5 and 6; Robert W. Herdt and A. M. Mandac, "Modern Technology and Economic Efficiency of Philippine Rice Farmers," Economic Development and Cultural Change 29 (1981):379-399; David W. Norman, "Economic Rationality of Traditional Hausa Dryland Farmers in the North of Nigeria" in Tradition and Dynamics in Small Farm Agriculture: Economic Studies in Asia, Africa, and Latin America, ed. Robert D. Stevens (Ames: Iowa State University Press, 1977), pp. 63-91.

[7]Hayami and Ruttan, Agricultural Development, especially chapter 6.

[8]Robert E. Evenson and Hans P. Binswanger, "Technology Transfer and Research Resource Allocation," Induced Innovation: Institutions and Development, eds. Hans P. Binswanger and Vernon W. Ruttan (Baltimore: The Johns Hopkins Press, 1978).

[9]For discussion of the costs of technology transfer see Evenson and Binswanger, Induced Innovation, pp. 167-169.

[10]See Yotopolous and Nugent, Economics of Development.

[11]The fairly large literature on augmenting agricultural research at the design and capacity transfer stages include additional discussion of skill levels at different phases in the transfer of technology. See Robert E. Evenson, "The Organization of Research to Improve Crops and Animals in Low-Income Countries" in Distortions of Agricultural Incentives, ed. by Theodore W. Schultz (Bloomington: Indiana University Press, 1978), pp. 240-242, and also James K. Boyce and Robert E. Evenson, National and International Agricultural Research and Extension Programs (New York: Agricultural Development Council, 1975), pp. 78-100.

[12]See for example Amir U. Khan, "Appropriate Technologies: Do We Transfer, Adapt or Develop?" in Edgar O. Edwards, ed. Employment in Developing Nations, (New York: Columbia University Press, 1974), pp. 223-234.

[13]An exploration of the range of distortions in prices in developing nations is provided in Distortions of Agricultural Incentives, ed. Theodore W. Shultz (Bloomington, Indiana: Indiana University Press, 1978).

[14]Induced institutional development is presented in more detail by Vernon W. Ruttan, "Induced Institutional Change," in Induced Innovation: Institutions and Development, eds. Hans P. Binswanger and Vernon W. Ruttan (Baltimore: The Johns Hopkins Press, 1978), Chapter 12.

[15]Theodore W. Schultz, "A Policy to Redistribute Losses from Economic Progress," Journal of Farm Economics 43 (1961): 554-565.

7
Risks and Information: Farm Level Impediments to Transforming Traditional Agriculture

Terry Roe

The body of evidence accumulated over the last decade generally supports Schultz's hypothesis of many years ago that "there are comparatively few significant inefficiencies in the allocation of factors and production in traditional agriculture." It also appears that farmers in traditional agriculture are fairly quick to adjust to the use of new methods of production if their welfare can be enhanced as a consequence. Surveying the literature on the adoption of food production technology, Ruttan concluded that the new wheat and rice varieties were adopted at exceptionally high rates in those areas where they were technically and economically superior to local varieties.[1] Perrin and Winkelman's review of CIMMYT sponsored studies on the adoption of wheat and maize varieties and fertilizer use arrived at a similar conclusion.[2]

The evidence implies that significant increases in resource productivity must come from the increased use of chemical, biological and mechanical technology, i.e., manufactured inputs. However, studies of traditional agriculture also reveal that farmers tend to be risk averse, and that returns to resources tend to vary depending on farmers' experience, cognitive ability, and access to information.

This chapter focuses on these factors as impediments to transforming traditional agriculture from a situation where farmers supply the bulk of their own inputs to a situation where they employ manufactured inputs. The household dimension of the transformation process, though important, is basically left to a future paper.

The chapter is organized by first discussing how farmers' perceptions (understanding) of the production characteristics of manufactured inputs, their cognitive ability, and access to information can give rise to resistance or impediments to substituting new for traditional technologies. Then, the conceptual approach and results from studies of Tunisian wheat and Thai rice farmers are presented to provide further--though still limited--insights into these impediments. Implications of these results to transforming traditional agriculture conclude the paper.

RISK AND INFORMATION AS IMPEDIMENTS TO
TRANSFORMING TRADITIONAL AGRICULTURE

It can be inferred from the studies of Binswanger,
Grisley, Dillion and Scandizzo, Roe and Nygaard, to mention a
few, that farmers in developing countries are almost always
risk averse.[3] That is, farmers experience disutility from not
knowing future income levels with certainty. Consequently,
they are willing to accept lower resource returns, and therefore lower incomes, if the certainty of future income can be
enhanced. Depending on opportunities to diversify or otherwise
share risk with other sectors of the economy, risk can be
viewed as a source of inefficiency in the allocation of resources. The level of risk aversion in some of these studies was
found to vary depending on the wealth of the farmer, his age
and education. Income variability in traditional agriculture
is most frequently attributed to yield and price variability.
Little attention has been given to health, nutrition and off-
farm income as sources of income variability.
    Several studies following the tradition of Nelson and
Phelps and Welch's research on the effect of education on resource productivity, have also shown that resource allocation
efficiency among farmers in traditional agriculture is associated with their cognitive ability and in some cases, access to
extension services.[4] Education is presumed to enhance farmers'
ability to receive, decode, and understand information. Hence,
education can become especially important if efficiency gains
are to be realized from the introduction of new technologies
into traditional agriculture. Education is also felt to affect
the "quickness" with which the maximum productivity gains can
be realized from adapting new inputs to a farmer's particular
situation.[5]
    Therefore, cognitive ability, risk attitudes and informa-
tion accessibility give rise to three potentially important
impediments within the farm-firm itself that remain to be overcome if the use of and efficiency gains from new technologies
are to be realized. The first impediment is the body of information that a farmer must assimilate prior to a decision of
whether to reject or adopt new technology(s). This process was
recognized and described by Beal, Rogers and Bohlen.[6] Only
recently have studies, e.g., Dryan and Lindner, attempted to
analytically treat this problem as one of information flows to
"update" previously held beliefs.[7] This impediment appears as
a time lag over the interval from when the technology becomes
available to when it is adopted. During this interval, of
course, efficiency gains are foregone while the farmer is evaluating information to determine whether or not to adopt the new
technologies. This is particularly true for new seed varieties
which tend to be site specific.
    The second impediment is the farmers' ability to correctly
perceive the production characteristics of a new technology.

86

In the case of new seed varieties, for instance, characteristics include the variety's yield response to fertilizer, tillage, land quality, pests, weather, etc. The efficiency gains realized from a technology are, of course, dependent on the extent to which the characteristics of the technology are correctly perceived. This is a cognitive process. As such, it can be influenced by formal education, experience, extension activities and the content and cost of information, including the resources the farmer allocates for experimentation purposes.[8]

The difficulty of correctly perceiving production characteristics also depends on the extent to which these characteristics depart from those of traditional technologies with which the farmer has had previous experience. Many technologies increase man's potential for controlling the vagaries of nature. Control is often accomplished by increasing the number and/or responsiveness of control variables or decreasing yield sensitivity to non-control factors such as disease and pests. The increase in control capability, however, also presents the potential for making allocative errors. Allocative errors, the potential of which are certainly of concern to farmers, decreases the efficiency gains that are possible from new technologies. As a corollary, the potential for allocative errors gives rise to the value of information and cognitive ability.

The third impediment relates to the risk averse attitudes of most farmers in traditional agriculture. These attitudes are essentially not, nor is it suggested that they should be, subject to change. Nevertheless, risk averse attitudes can impede the realization of efficiency gains from new technologies in at least two different situations.

Suppose, for purposes here, that farmers' perceptions of production characteristics are described by some multivariate probability distribution. Then, the degree of confidence farmers have in their perceptions can be measured by their subjective probabilities that these characteristics can take on different values. Now, given any level of confidence in their perceptions, i.e., a probability distribution, the efficiency gains obtained from the technology will depend on the farmer's risk attitudes. To illustrate, two producers may have identical beliefs that the employment of a new technology will yield higher returns than traditional alternatives. Yet, the more risk averse producer may not employ, or only employ low levels of the technology. In a Bayesian framework, it can be shown that higher levels of risk aversion also impede the farmer's learning of the production characteristics of new technology and, hence, prolong the attaining of the maximum efficiency gains from new technologies.

The second way in which risk attitudes affect the use of new technologies relates to the financial commitment the use of these technologies imply. The use of new technologies almost always implies a larger financial commitment by producers prior to the realization of returns at harvest.[9] Given any level of confidence in perceptions, the more risk averse producers may be reluctant to adopt or make any extensive use of the technology, thereby foregoing any efficiency gains that might be realized.

These impediments have been discussed within the context of the farm as a firm. In traditional agriculture, the farm is a firm-household. The so-called New Household Economics has been applied in LDC's by a number of writers.[10] From the theory of the agricultural household it can be shown that household endowments, household commitments on the production of food by the farm-firm and preferences for the allocation of labor and savings can influence the impediments mentioned above (in part because the producers' risk preferences are directly related to the household's demand for goods and services). It can also give rise to other impediments. For instance, this theory suggests that the uncertainties in capital and labor markets and the magnitude of product marketing margins for goods consumed by the household decrease incentives to use new technologies. However, empirical research on this topic is only beginning. Hence, it is perhaps to soon to draw inferences from the theory itself.

Recent farm level studies of wheat and rice production in Tunisia and Thailand respectively provide further insight into the nature of, particularly, impediments two and three. A brief description of the approach and selected results of these studies are presented in the next section.

THE EFFECT OF RISK AND INFORMATION ON WHEAT AND RICE PRODUCTION IN TUNISIA AND THAILAND

Conceptual Framework

The conceptual approach employed in the Tunisian and Thailand studies are virtually identical. The maintained hypothesis underlining the analyses is that farmers allocate resources in order to maximize their expected utility of gains. More formally, the problem an individual farmer faces is to commit input levels X_k to maximize

$$E[U_n] = E[U(\pi_n)]$$

where $E[\cdot]$ is the expectations operator and $U(\pi_n)$ is the utility of gains function for the n-th producer. The expected utility of gains can be viewed as an indirect function of utility for goods and services of the household. It is influenced by household endowments and other household characteristics.

The key to the analysis lies in the specification of the gains function π_n.

First, let

(1) $Y = f(X;m)\epsilon$

denote the true physical correspondence between output (Y) and a K element vector of K* choice and K-k* nonchoice inputs where m is a vector of parameters and ϵ is a disturbance term. This function can be thought of as the production function that is often estimated in farm level studies. The important question is whether the parameters (m and ϵ) of (1) are known and whether the levels of the K-k* nonchoice variables can be assessed by farmers.

Since farmers must first make resource commitments if production is to take place, we assume, and later empirically test, that producers formulate a subjective density on the value of the parameters (m, ϵ) and noncontrolled variables in (1). Let

(2) $Y_n^p = f(X;m_n)v_n$

denote this subjective (or behavioral) production function. The n-th producer's subjective estimates of the parameters in (1) are given by m_n, and v_n. These "subjective" parameters may in turn depend on cognitive ability, experience and access to information and extension services.

The gains function (π_n) is now defined as the gains the farmer <u>expects</u> to obtain at harvest, conditional on input choices he makes at, e.g., the beginning of the production period:

(3) $E[\pi_n] = PE[f(X;m_n)v_n - \sum_{k}^{K*} P_k X_{kn}$

Note that instead of the "true" production function (1) the subjective function (2) determines expected gains. Hence, the farmer's choice of inputs will depend on his "perception" of the true production relationship rather than on what the relationship actually is.

If farmers' estimates are not accurate, and/or if they behave as though their estimates, m_n, and v_n, have some distribution about the true parameters, m, ϵ of the technology, then uncertainty and allocative errors can occur.

Under various conditions, it can be shown that the expected utility function $E[U(\pi_n)]$ can be expressed as a function of expected gains (3) and the variance of gains $[V\pi_n]$. Then the maximization of expected utility yields the condition:

(4) $\Phi_n \partial V[\pi_n]/\partial X_{kn} = PE[f(X;m_n)v_n]/X_{kn} - P_k$

where it has been shown by others that

$$\Phi = \{- \frac{\partial E[U]}{\partial V[\pi_n]} / \frac{\partial E[U_n]}{\partial E[\pi_n]} = 0\} \quad \begin{array}{l} \text{risk averse} \\ \text{risk neutral} \\ \text{risk preferred} \end{array}$$

The parameter Φ is referred to as the risk evaluation differential quotient.[11] If a producer is risk averse, and if the marginal variance, $\partial V[\pi_n]/\partial X_{kn}$, is positive then the producer does not allocate inputs to the point where the expected return from the last unit of the resource is just equal to its price.[12] That is, he is reluctant to obtain all possible net returns from his inputs because of the chance or risk that some unexpected event could occur and result in lower than expected yields and possible economic losses.

Several alternative measures can be used to estimate the costs and/or returns foregone if (1) differs from (2). One measure, the value of perfect information, is the value to the farmer of knowing the true production characteristics exactly. The value of perfect information is given by

$$E[\pi_n^*] - E[\pi_n] \geq 0.$$

The expected value of π_n^* is determined by the level of the choice variables a farmer would have chosen if (1) were known with certainty. The expected value of profits realized, π_n^0, is determined by the farmers' input choices, while the yields expected to be realized at harvest are given by.[13] A diagram of the relationship between these functions appears in Figure 7.1. A similar relationship can be shown to exist for the cost function, except in this case, the cost function realized in their forecast was also solicited in order to provide estimates of the variance $V[\pi_n]$. Data were also collected on the yields obtained at harvest, farming experience, age, education, access to extension agents and other sources of information. The results of fitting the framework discussed above to data for the case of wheat and rice production in Tunisia and Thailand, respectively, are discussed in the next section.

Empirical Results.

Farmers' yield forecasts and the yields they actually obtained at harvest are reported in Table 7.1. Overall, farmers in both countries expected to obtain higher yields from the high yielding varieties than they actually obtained. In both cases, their yield forecast of the high yielding varieties was more inaccurate than their forecast of the local varieties. This should not be surprising since they are more familiar with growing the local varieties.

Figure 7.1. Relationship Between Subjective $E(\pi_n^0)$, True $E(\pi^\star)$ and Realized $E(\pi_n)$ Profits for the Case Where $E(Y_n) \geq E(Y)$ for all X_1^0, X_2^0.

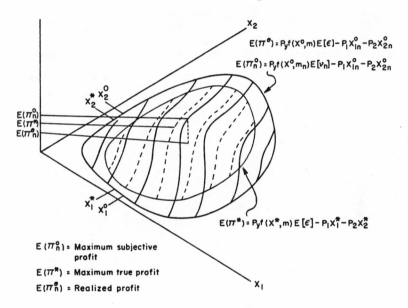

$E(\pi^0) = P_y f(X^0, m) E[\epsilon] - P_1 X_{1n}^0 - P_2 X_{2n}^0$

$E(\pi_n^0) = P_y f(X^0, m_n) E[\nu_n] - P_1 X_{1n}^0 - P_2 X_{2n}^0$

$E(\pi^\star) = P_y f(X^\star, m) E[\epsilon] - P_1 X_1^\star - P_2 X_2^\star$

$E(\pi_n^0) = $ Maximum subjective profit

$E(\pi^\star) = $ Maximum true profit

$E(\pi_n^0) = $ Realized profit

91

TABLE 7.1

Average Yields Obtained at Harvest and Yields Expected at the Time of Seedbed Preparation, for Wheat and Rice in Tunisia and Thailand*

| | Tunisia | | Thailand | |
| | Mean Wheat Yield in Quintals/Ha. | | Mean Rice Yields in Tang/rai | |
	Observed 1976/77	Forecast 1976/77	Observed 1980	Forecast 1980
HYV	7.51	13.20	58.32	66.78
LV	5.44	8.49	50.13	48.34

*Based on farm level survey.

The next step is to identify the source of this error. This is accomplished by estimating and comparing the true function (1) with the subjective function (2). The results for Tunisia appear in Table 2. Coefficients of the true function for high yielding varieties and the true function for local varieties appear in the first two columns of the table. Estimates of the subjective function (2) appear in the third column. Only a single column of parameter estimates are reported for the subjective function because statistical results from fitting the subjective function for each of the varieties indicated that the equations only differed in the intercept term. Hence, the equations were combined and reestimated. This implies that at the time of seedbed preparation in 1976, Tunisian farmers acted as though there was no difference between varieties in response to the inputs they control.

Before the Tunisian results are discussed, it is important to point out the unusual nature of the 1976/77 growing season. When farmers' subjective forecasts were solicited, soil moisture was ample. However, during the remainder of the growing season, the winter rains arrived late so that soil moisture became unusually scarce. This is in part revealed by the magnitude of the estimated intercept term (1.3882) in the subjective function. Results reported by Gafsi and Roe for the 1972/73 crop year also permitted the combining of the high yielding and ordinary yielding equations into a single equation.[14] The intercept term reported in those results was 1.323 and it was not significantly different from the intercept term in the subjective function in Table 2. Thus, for a normal year farmers are perhaps correct in visualizing that there is no individual input productivity difference between the high and ordinary yielding varieties except for a yield difference. It is likely, therefore, that the difference in slope coefficients reported here is in part explained by abnormal weather conditions of the 1976/77 crop year.

TABLE 7.2

Parameter Estimates of the True and Subjective Functions (1) and (2) for the 1976/77 Crop Year, Northern Tunisia

Variables	Equation (1) HV True		Equation (1) OV True		Equation (2) Subjective Estimates	
B = constant term	.5425	(2.1)	.7596	(4.8)	1.3882	(17.3)
D_1= HYV					.3604	(6.2)
D_2= soil	.3712	(3.1)	.3959	(2.9)	.2966	(5.7)
D_3= zone	-.3887	(3.3)	-.3987	(2.3)	.1577	(3.2)
Ex = 1/yr of experience					-.2054	(2.3)
Ph = phosphate	.1525	(3.2)	.1031	(2.4)	.0406	(2.3)
N = nitrogen	.0163	(0.4)	-.0134	(0.3)	.0645	(3.7)
M = machinery	.3375	(4.0)	.1856	(3.0)	.1063	(3.7)
L = land	.3718	(3.0)	.7874	(7.6)	.8301	(18.6)
R^2	77.0		79.0		93.2	
n	127		98		228	
F	72.5		61.06		388.44	

All coefficients are significant at the 99 percent level except the nitrogen coefficients.

Source: Roe and Nygaard.

The dummy variable (D1) suggests that at the time of seed-bed preparation Tunisian farmers expected the high yielding varieties to out-yield the ordinary yielding varieties by at least 30 percent. The coefficient on the soil dummy (D2) implies that farmers expected that wheat planted on good soil will also increase its yield in the vicinity of 30 percent. The negative sign on the experience variable suggests that as farmers gain more experience with the variety, they expect to obtain higher yields, but the rate at which they expect these yields to increase, decreases with each additional year of experience.

A comparison of the differences in the slope coefficients suggest that, based on a t-test, farmers underestimated productivity of phosphate fertilizer and overestimated the productivity of nitrogen fertilizer; a comparison of the machinery coefficient suggests that they also underestimate the productivity of machinery. The relatively large coefficient on land is puzzling except that they are of approximately equal magnitudes.

The results for rice in Thailand appear in Table 7.3. In this case, the differences between the two functions (1) and (2) are not as striking as in the Tunisian case. Perhaps this is not surprising since rice is grown under irrigated conditions which implies more control than wheat grown under rain-fed conditions. The statistically different coefficients are in the intercept term and the two slope coefficients associated with broadcasting techniques and labor.

The smaller intercept term of the subjective function suggests that the farmers may have underestimated the scale effect of the rice technology, thereby "underallocating" inputs. In the case of seeding techniques, the results suggest that, as a group, Thai farmers believe that the broadcasting technique has no real effect on yields. Estimates of the "true" function suggest instead that broadcasting techniques tend to lower yields by about 2.7 percent. Thai farmers also appear to feel that labor is more productive than it actually may be.

Thai farmers' forecasts seem to be related to the amount of labor and farmers' age. Within the range of data, as age increases, the yield forecast converges to the mean yield observed at harvest. Farmers experience in growing high fielding varieties did not seem to have any affect on their forecasts. Otherwise, all of the remaining coefficients have the same sign and they are not significantly different at acceptable confidence levels. Hence, as a group, their insights into the production characteristics of these varieties seem reasonably accurate.

If the conceptual framework outlined above is an accurate description of farmers' decision-making problems, then we would expect that these results should show that Tunisian farmers made larger errors in the allocation inputs than did Thai farmers.

The next step is to obtain insights into farmers' risk

TABLE 7.3

The Estimated Perceived and the Estimated True Production Functions per Rai, Thailand, 1980

Variables	OLS		SUR	
	True	Perceived	True	Perceived
Constant	1.0337	0.6917	1.0337	0.5941
	(1.6152)	(1.4648)	(1.6152)	(1.2610)
X_1 = % of Area with HV	0.1687	0.3460***	(1.6152)	(0.3028)***
	(0.6462)	(1.6570)	(0.6462)	(1.6473)
X_2 = % of broadcast area	-0.0273*	0.0100	-0.273*	0.0097
	(-3.5361)	(1.7869)	(-3.5361)	(1.7245)
X_3 = Labor	0.1491*	0.3974*	0.1491*	0.3887*
	(2.6299)	(8.5057)	(2.6299)	(8.4418
X_3 . Age	--	-0.0018*	--	-0.0013*
		(-3.9268)		(-3.0400)
X_3 . Experience	--	0.0024	--	-0.0071
		(0.1027)		(-0.3201)
X_4 = Machinery	0.3524*	0.2033**	0.3524*	0.2135*
	(3.5999)	(2.8467)	(3.5999)	(2.9910)
X_5 = Insecticide	0.0416	0.0634**	0.0416	0.0651**
	(1.0555)	(2.2030)	(1.0555)	(2.2642)
X_7 = Fertilizer	0.1385	0.1448*	0.1385	0.1499*
	(1.8334)	(2.5878)	(1.8334)	(2.6866)
X_7 . X_1	0.0477	-0.0455	0.0477	-0.0409
	(0.4479)	(-0.4888)	(0.4479)	(-0.5298)
X_7 . Number of plots	-0.0416**	-0.0309**	-0.0416**	-0.0332**
	(-2.1393)	(-2.0940)	(-2.1393)	(-2.2561)
X_7 . X_2	-0.1167**	-0.472	-0.1167**	-0.0507
	(-2.0255)	(-1.1038)	(-2.0255)	(-1.1893)
R^2	43. 65	60.47	43.65	60.30
F	11.53	18.36		
n = 142				

*Significant at 99%, ** significant at 95%, *** significant at 90%. OLS and SUR denotes ordinary least squares and seemingly unrelated regression estimates respectively.

Source: Somnuk Tubpun and Terry Roe.

95

attitudes. These estimates are obtained by substituting the estimates of (2) into the right hand side of (4) and then deriving the left hand side of (4). The risk evaluation differential quotient, Φ, is obtained by computing the marginal risk $\partial V[\pi_n]/\partial X_{nk}$, in the Tunisian study, and $\partial V[\pi_n]/\partial Y_n^p$, in the Thailand study from the survey data. The means of these values over the farmers sampled are reported in Table 7.4.

Table 7.4

Summary of Risk Aversion Estimates for Wheat and Rice Products in Tunisia and Thailand, Respectively

	Tunisia		Thailand	
	Φ	$\Phi\partial V(\pi_n)\partial X_{nk}$	Φ	$\Phi\partial V(\pi_n)/\partial Y$
	Dinars/Qx		Baht/tang	
Mean	.003418	1.851	.0011	13.197
Std. Deviation	.011942	3.661	.0015	17.408

These values must be interpreted with caution since credit or other input constraints that may exist are ignored. If these constraints affect the farmers choice of inputs, then the estimates of the affects of risk are almost surely biased upwards. Nevertheless, when compared to other studies, Φ falls within a reasonable range. In the case of Tunisian, 73 percent of the farmers in the sample were risk averse, 19 percent were risk neutral, and 8 were risk preferring. Virtually all of the farmers in the Thailand study are risk averse with $0 < \Phi < .0015$ for about 114 farmers (79 percent).

The average magnitude of the risk discount for Tunisian farmers is about 1.85 dinars per quintal. In the case of Thailand, the discount is about 13.197 bahts per tang. These discounts are about 25 percent of the price of a quintal of durum wheat in Tunisia and about 17 percent of a tang of paddy rice in Thailand. Tunisian farmers' perception of risk in growing the high yielding varieties exceeded their perception of risk in growing the ordinary varieties. The standard deviation of expected profit per hectare of the high yielding varieties exceeded that of ordinary varieties by about 27 percent. This accounts in part for the larger percentage discount for risk in Tunisia relative to that in Thailand.

In both studies, the estimated value Φ for each farm was regressed on a number of producer and farm characteristics. In the Tunisian case, the farmers' age and whether the farm was located on hilly or valley land were found to be positively correlated with farmers' risk attitudes. The land area farmed was negatively correlated with farmers' risk attitudes. A negative correlation suggests that, if farm size is proxy for wealth, as wealth increases farmers aversion to risk decreases. The negative correlation between farmers' risk attitudes and

farm size was also found in the Thailand study. As the age of
farmers in Thailand increased, their risk aversion decreased.
Education and the number of extension agent contacts were found
to be positively correlated with farmers aversion to risk in
Thailand. These results suggest that farmers, as a group, tend
to be "cautious" and, under conditions of positive marginal
risk, allocate fewer resources than they would with risk neutral attitudes.

The production function results showed that differences
existed between the "true" production function (1) and farmers
estimates of this function. Equation (1) can be used to com-
pute the least cost combination of variable inputs for the
level of output farmers expected to realize at harvest. A comparison of these costs with the costs farmers actually incurred
provides insights into the efficiency with which resources were
allocated.

As a group, Tunisian farmers actually incurred, on average, a cost per unit of wheat produced at about two times the
level of the least cost they would have incurred had they
known, at the time of seedbed preparation, the true production
function for HYV's. In other words, for yield levels of the
high yield varieties realized at harvest, input levels could
have been reduced and their relative proportions changed so
that unit costs would have fallen to about half of the unit
costs actually incurred. In the case of ordinary varieties,
unit costs exceeded least costs at yields realized by a multiple of about 1.6. While larger, errors in this case were
clearly less than in the case of the high yielding varieties.

The abnormal year clearly contributed to these larger
errors. Had weather conditions for the 1976/77 crop year been
similar to the 1972/73 crop year, then the production function
estimated by Gafsi and Roe suggests that allocative errors on
average would have exceeded least cost by about 25 percent. It
is also worthwhile to note that the more risk averse farmers
suffered smaller losses because their input levels were lower
than input levels of the risk averse farmers.

In the Thailand study, the level of allocative error
appears much smaller. A comparison of the variable costs per
unit of output realized with least costs indicates that, on
average, farmers in Thailand only exceeded least costs by about
16 percent. Again, these results must be interpreted with caution. The analysis ignores several factors, among which are
access to inputs, assumptions underlying expected utility as a
function of mean and variance, and possible statistical biases
inherent in the production function estimates.

In the Thailand study, the value of perfect information
was computed based on condition (5). The value of information
was defined as the difference between the maximum attainable
net returns per rai (a measure of area), based on known prices,
and the level of observed net returns per rai realized at harvest. Under these restrictive assumptions, returns over vari-

97

able costs could have been increased by about 28 percent on the average. This estimate is very likely biased upward because, under conditions of perfect information, farmers are assumed to face no risk. This is unlikely to occur in reality. Nevertheless, it provides an upper bound to efficiency gains that might be attainable from extension activities.

The next and final question here is whether the proportion of land planted to high yielding varieties is affected by farmers risk attitudes and other farm level characteristics. These results are reported in Table 7.5 for Tunisia and Table 7.6 for Thailand.

In the Tunisian case, the signs of the explanatory variable conform with expectations. However, only 27 percent of the variation in the dependent variable is explained by the independent variables. The results show that risk attitudes are a major deterrent to increasing the area planted to high yielding varieties in Tunisia. The results also show that valley land is preferred to hilly land for growing high yielding varieties. Farm size, extension agent contacts, and education all appear to have positive effects. However, the coefficients associated with these variables are not significantly different from zero at acceptable confidence levels.

In the case of Thailand, both a probit and ordinary least squares estimating procedures were used. Both equations appear to fit the data fairly well; the results are much improved relative to the Tunisian case. While risk appears as a significant explanatory variable in the case of Tunisia, it has alternating signs in the tobit and ordinary least squares results. Hence, the risk attitudes of Thai farmers appear not to be significant factors in determining the area planted to high yielding varieties of rice.

Farmers whose land was located close to the head of the irrigation canals seem to plant a higher proportion of their land to high yielding varieties than those at the end of the canals. This is felt to be attributed to the case of water management. Years of experience with high yielding varieties also has a positive effect on the rate of adoption. On the other hand, higher farm level prices of local varieties and larger numbers of plots in each farm inhibit the adoption of high yielding varieties. Experience with growing high yielding varieties and whether high yielding varieties were grown on a farm in previous years also appear to have a positive effect on the area planted.

IMPLICATIONS

The extent to which our theoretical devices reflect the decision making process and the extent to which these devices fit the survey data determines the validity of the conclusions drawn. Since this area of inquiry is complex, additional research is required to confirm or extend these results. Yet,

TABLE 7.5

Adoption of High Yield Varieties as a Function of Socio-Economic Variables, Tunisia, 1976/77

Variables	Regression Coefficients	
A = constant term	-2.3536	(4.09)**
= risk parameter	-27.5069	(2.07)*
FS = farm size	.0049	(1.61)
Ext = extension contact	.0385	(0.89)
Ed = education	.0015	(0.04)
T = topography	1.1396	(2.39)**
Z = zone	1.1446	(1.14)
R^2	.27	
f	4.36	
n	78	

*Significant at the 97.5 percent level.
**Significant at the 99 percent level.

Dependent variable is in $\dfrac{\text{area HYV of durum wheat}}{\text{total area of durum wheat}}$

Topography = dummy variable where 1 = valley land
0 = hilly land

Zone = dummy variable which is the same as in production function estimates

t values are in parentheses

99

TABLE 7.6

Adoption of High Yield Varieties as a Function of Socio-Economic Variables, Thailand, 1980

Explanatory Variables	Description	Fitting the models by	
		Tobit	OLS
Constant Term		4.6046* (5.113)	2.8089* (4.3403)
	Degree of risk aversion	-13.7960 (-0.513)	0.4501 (0.0217)
M	Location of farm land	0.2097* (2.648)	0.1739* (3.4304)
1/EX	Years of experience with HV	0.1558* (5.807)	0.0826* (4.6087)
LX_6	Last year area planted to HV	0.214* (4.051)	0.0151* (4.4055)
X_6	This year farm size	-0.0078*** (-1.749)	-0.0055** (-1.9709)
NPT	Number of fragments of farm land	-0.0835** (-2.2405)	-0.0555** (-2.2405)
Z_2	Need of production loans	-0.1243*** (-.1750)	-0.0725 (-1.6065)
FP	Farm level prices of LV	-0.1533* (-5.474)	-0.0831* (-4.7240)
ME	Number of household members	0.0373*** (1.747)	0.0259*** (1.8970)
AG	Farmers age	-0.0050*** (-1.800)	-0.0028 (-1.5384)
	R^2 =	64.28	R^2 = 60.93
	-2 log =	137.25	F = 17.02

Pr[y>0] given average explanatory variables .6523
y at means of average explanatory variables .2192

*Significant at 99 percent confidence level; **Significant at 95 percent confidence level; ***Significant at 90 percent confidence level.

Dependent variable $\dfrac{\text{area HYV of rice}}{\text{total area of rice}}$

Source: Somnuk Tubpun and Terry Roe.

100

the Tunisian and Thailand studies integrate, in a single framework, the effects of both risks and farmers' knowledge of production characteristics on the efficiency of resource use.

The studies of Tunisian and Thai farmers, though limited, provide further evidence in support of risk and information as impediments in transforming traditional agriculture from the use of traditional farmer supplied inputs to the use of manufactured inputs. The findings suggest that farmers' understanding of production characteristics of technologies available to them are important determinants of their input choices and the productivity realized from these choices. In the case of Tunisia, allocative errors were largely because of the unpredictable and uncontrollable effect of weather on yields. Even if expected weather conditions had prevailed, incorrect perceptions of the responsiveness of the yields of high yielding varieties of durum wheat would have caused allocative error in the vicinity of about 25 percent of least cost based on knowledge of the "true" production function.

In the case of Thailand, farmers appeared to more accurately perceive the production relationships they actually faced. Accordingly, estimates of allocative error suggested that realized unit costs only exceeded least costs by about 16 percent. Hence, 16 percent is the maximum gain per unit, at production levels realized in 1980, that extension activity could hope to create from informing farmers of the production characteristics of the rice varieties grown. However, because of risk attitudes, if this information also served to remove risks completely, then farmers would also increase production. In this case, the maximum gain in net revenue would be in the vicinity of about 28 percent.

In Tunisia, farmers' risk attitudes were important determinants of the area planted to high yielding varieties of durum wheat. While Thai farmers also appear to be risk averse, their degree of risk aversion was somewhat less than that of Tunisian farmers. Their risk attitudes appeared to have little effect on the area they planted to high yielding varieties of rice. Evidence also was found to suggest that education, age and experience influenced the use of high yielding varieties.

Other inferences from the results are that trade-offs in efficiency gains realized likely exist between the extent to which a technology differs from traditional technologies (i.e., its relative complexity) and the potential gains from a new technology. Often, higher potential inefficiency gains are associated with greater relative complexity of the technology. In an environment where it is difficult to acquaint farmers with the characteristics of a new technology, then the efficiency gains actually realized may be higher than with a less complex technology which offers lower potential for efficiency gains. In the case of a relatively less complex technology, efficiency gains should be realized in a relatively shorter period of time.

101

Another dimension to the characteristics of new technologies is the production control they permit farmers to exercise during the growing season. In the case of irrigation, Thai farmers clearly faced a more predictable and stable environment than did Tunisian farmers, not only because of water control but also because inputs, fertilizer and other chemicals could be added during the growing season. In the case of dry land wheat production, the majority of the inputs are committed during the early stages of the growing season. The inability to adjust input levels as the accuracy of farmers forecast increase with the approach of harvest, tends to increase risk and decrease the attaining of the maximum efficiency gains from a technology. Technologies that increase the duration of control over the growing season should also increase the likelihood of attaining the potential efficiency gains from a new technology.

All else equal, the greater the degree to which the technology differs from farmers' previous experiences, the more resources that are likely to be required to acquaint the farmers with the new technology. In this case, extension programs will be more successful when they empathize with farmers' decision making problems, and take into consideration the farmers' present state of knowledge and beliefs regarding the production possibilities of new technologies and the yield variability the new technology implies under various weather conditions.

Technologies that differ in important ways from older familiar technologies, may tend to be adopted more quickly on larger farms. Larger sized farms can permit more risk diversification. They can probably afford to experiment and learn the peculiarities of the new technologies in a shorter period of time than is optimal for smaller sized farms. In India, Sidhu found the varieties to be neutral in the use of inputs, and quickly adopted by farmers.[15]

In Tunisia, farmers' risk attitudes were important determinants of the area planted to high yielding varieties of durum wheat. While Thai farmers also appear to be risk averse, their degree of risk aversion was somewhat less than Tunisian farmers. Their risk attitudes appeared to have little effect on the area they planted to high yielding varieties of rice. Evidence also was found to suggest that education, age and experience influenced the use of yielding varieties. It was also found that most of the high yielding soft wheats were grown on larger farms. Hence, unless efforts are made to subsidize the learning process of farmers on smaller sized farms, income distribution inequalities can result.

In the case of new and unfamiliar technologies, the farmers' perception of risk, as indicated by the empirical results, likely exceeds the risk faced with the old technology. This suggests that additional consideration should be given to the design of farm level programs which, while not distorting price signals, permit farmers to lower risk in the early stages of the adoption process. One such program may be a subsidized crop insurance program. This type of program can be socially profitable in the short run if the program results in a rate of increase in efficiency gains that exceeds the cost of the program. However, crop insurance programs can become socially unprofitable if the gains to society are lost to the cost of administration.

If growth in agricultural production is to be achieved through increased use of purchased inputs, the structure and stability of input and product markets become even more important. Furthermore, as has been pointed out by Schuh and others, over-valued currency exchange rates and cheap food policies for the urban masses can amount to a substantial tax on the agricultural sector.[16] Producer investments in information in order to become acquainted with new technologies will, like other investments, almost surely be discouraged if the agricultural sector is discriminated against by other macro policies.

REFERENCES

[1]Vernon Ruttan, "The Green Revolution: Seven Generalizations," International Development Review 19 (1977):9-20.

[2]Richard Perrin and Don Winkelmann, "Impediments to Technological Progress on Small Versus Large Farms," American Journal of Agricultural Economics 58 (1976): 888-894.

[3]Hans P. Binswanger, "Attitudes Toward Risk: Experimental Measurement in Rutal India," American Journal of Agricultural Economics 62 (1980):395-407; William Grisley, Effect of Risk and Risk Aversion on Farm Decision-Making: Farmers in Northern Thailand (Ph.D. thesis, University of Illinois, 1980); John L. Dillion and Pasquale L. Scandizzo, "Risk Attitudes of Subsistence Farmers in Northern Brazil: A Sampling Approach," American Journal of Agricultural Economics 60 (1978): 425-435; Terry Roe and David Nygaard, Wheat, Allocative Error and Risk: Northern Tunisia, Bulletin V (Economic Development Center, Department of Economics and Department of Agricultural and Applied Economics, University of Minnesota, 1981).

[4]Richard Nelson and Edmund S. Phelps, "Investments in Humans, Technological Diffusion and Economic Growth," American Economic Review 56 (1968): 59-75; Finis Welch, "Education in Production," Journal of Political Economy 78 (1970): 35-39; Marlaine E. Lockheed, Dean J. Jannesson and Lawrence J. Lau, "Farm Education and Farm Efficiency: A Survey, "Economic Development and Cultural Change. The latter is a review of research from countries in Asia, Africa and Latin America showing that a large number of studies reported a positive and unusually significant contribution of education and extension to resource productivity.

[5]Dayanatha Jha and Rakesh Sarin, An Analysis of Levels, Patterns and Determinants of Fertilizer Use on Farms in Selected Regions of Semi-Arid Tropical India (Andhra Pradesh, India: International Crops Research Institute for the Semi-Arid Tropics, Oct. 1981). Their regression results from India also show that the determinants of fertilizer use included the level of knowledge of the farmer as measured by experience and education. Risk aversion was not found to be an important constraint everywhere--only in regions experiencing yield variability due to poor irrigation infrastructure or dry land farming.

[6]G. M. Beal, E. M. Rogers and J. M. Bohlen, "Validity of the Concept of Stages of the Adoption Process," Rural Sociology 22(1957): 166-168.

[7]Ross G. Drynan, "Experimentation--It's Value to the Farm Decision Market" (Ph.D. thesis, Armidale, N.S.W., 1977); Robert Lindner, Adoption as a Decision-Theoretic Process (Ph.D. thesis, University of Minnesota, 1981).

[8]Drynan, "Experimentation--It's Value to the Farm Decision Market."

[9]Hans P. Binswanger and Donald A. Sillers. "Risk Aversion and Credit Constraints in Farmer's Decision Making: A Reinterpretation" (Working Paper, IBRD, December 1981). A discussion of the relative roles of risk aversion and credit constraints in limiting farmers' investment levels, especially for purchased inputs such as seeds and fertilizers.

[10]Lawerence Lau, Wuu-Long Lin and Pan Yotopoulos, "The Linear Logarithmic Expenditure System: An Application to Consumption and Leisure Choice," Econometrica 48 (1978): 843-868; Choong Yong Ahn, Inderjit Singh and Lyn Squire, A Model of an Agricultural Household in a Multi-Crop Economy: The Case of Korea," Review of Economics and Statistics LXIII (1981): 520-525; John Strauss, Determinants of Food Consumption of a Household-Firm Model with Application of the Quadratic Expenditure System (Ph.D. thesis, Michigan State University, 1981).

[11]Gudmundur Magnusson, Production Under Risk: A Theoretical Study (Uppsala, Sweden: Almquist and Wiksells Boktryckeri AB, 1969).

[12]Richard E. Just and Rulon Pope, "Stochastic Specifications of Production Functions and Economic Implications," Journal of Econometrics 7 (1978): 68-86. In contrast to this analysis for the Cobb-Douglas case, the term can be positive or negative because m_n are subjective estimates of m.

[13]A procedure for estimating the value of information in the risk averse case is given by Frances Antonovitz and Terry Roe, A Measure of the Value of Information for the Competitive Firm Under Price Uncertainty (Paper contributed to Annual Meetings of the American Agricultural Economics Association, Logan, Utah, 1982).

[14]Salem Gafsi and Terry Roe, "Adoption of Unlike High-Yielding Wheat Varieties in Tunisia," Economic Development and Cultural Change 28 (1979): 119-133.

[15]Surjit S. Sidhu, "Economics of Technical Change in Wheat Production in the Indian Punjab," American Journal of Agricultural Economics 56 (1974); 221-232.

[16]G. Edward Schuh, Floating Exchange Rates, International Interdependence and Agricultural Policy (Proceedings, Seventeenth International Conference of Agricultural Economics, Banff, Canada, 1979).

8
Marketing Decisions for Small-scale Producers

Donald R. Street
Gregory M. Sullivan

INTRODUCTION

A strong linkage exists between marketing and production decisions in the world food industry. Marketing is strongly complementary with production precluding independent decision making in these two phases of an enterprise. The whole of the elements equals more than the sum of the parts. Risk and uncertainty spillovers from production to marketing and vice versa are mutually reinforcing. This chapter concentrates on the nature of the imperfections in the market structure for food producers and the implications for adjustments in the context of efficiency and equity.

Market imperfections probably exist to a much greater degree in the agricultural sector in underdeveloped countries than in developed countries. Adoption of appropriate production technology is constrained by many factors somewhat isolated from the production decisions per se. Adoption of appropriate technology in production and in marketing will also be constrained by imperfections in the market structure itself.

Policies formulated at the macro level may either preclude or open up decision-making opportunities at the micro level. This result is especially true of basic infrastructure at the country or regional level of developing areas. Improved efficiency at the macro level of marketing may also have serious redistribution effects among consumers or between consumers of products and sellers of products. This fact complicates policy planning by the question of who is the target group of aid projects. Aid to one group may be at the expense of another.

Government policy is the umbrella under which most individual marketing decisions are made. The stability of the government, its dependability, and its support are therefore paramount in establishing an efficient if not ideal setting for the marketing of food products. The government has control of the macro-setting which limits, modifies or proscribes micro-level decisions. The macro-setting either frees the system by opening up lanes to prosperity or enslaves micro decision makers by its lack of proper functioning.

106

The smooth and timely flow of products from producers to the final consumer is a goal of the marketing sector of the economy. There are numerous impediments to this smooth flow involving the functions of transport, storage, merchandising, financing, etc.

Infrastructure and technology transfer are not enough to solve the world food marketing problems. Matters of equity and market organization will be more important in the future in the whole marketing endeavor as links to attainment of national goals. Social and economic consideration for landless peasants and management of scarce resources are becoming of increasing importance for LDC's.

A large part of the authors' experience in marketing in LDC's has dealt with fisheries and especially with aquaculture. Several examples of marketing practices and problems will be examined from this perspective. Fish marketing principles could be especially useful in relation to other perishable products such as fruits and vegetables or meat and dairy products.

MARKETING SYSTEMS AND INFRASTRUCTURE

Modern-day marketing must be approached within a systems context in an effort to improve efficiency and minimize conflicts within a country. Decisions to produce, to purchase inputs, to store, to process, to transport, and to distribute at wholesale and retail levels must be coordinated. The horizon of decision making in the broadest context should fit into national goals and policies and should be compatible with national and regional planning objectives. The radius of influence of the supplies of products and their importance as a fraction of domestic and export goods will determine, in part, how much attention is given to specific commodities in the marketing program. Some products will have an influence only in the local area while others are critical in the sense that they are large enough in quantity and value for national concern either for domestic consumption or as a source of foreign exchange.

Marketing Systems
The producer must plan to set up his own marketing arrangement or look to some outside entity to do the job. The size of the production scale and the education level of the producers in most LDC's will necessitate that other agencies be involved in marketing. The institutions involved may be free market organizations, government buyers, cooperatives, or other entities such as quasi-government organizations which act as a framework for operation with private or public capital. There has been a great deal of resentment at individual and government levels toward various types of middlemen as buyers of produce at the farm level and at succeeding levels.[1] Appropriate

market institutions need to be designed to provide a proper functioning of the system.

The free-market, private-buying organization is probably the most common type of institution through which marketing is done in most of the free world. This group includes brokers and buyers of other types who may take title to the goods sold by producers. Some may resell at retail while others may sell only at wholesale.[2] Kamenidis has shown that there is a prevailing lack of esteem for this group of middleman types.[3] Kamenidis reports that "...Farmers' attitudes toward middlemen in general are often fueled by politicians seeking scapegoats for low prices to farmers or high prices to consumers."[4]

Obviously, the middlemen perform various functions and should be rewarded for the services they perform without accruing economic rents. Harrison points out that:

"There is a tendency to focus undue attention on all the real and imagined shortcomings of traditional middlemen; to disregard the need for the services performed by these market intermediaries; to overlook the costs and managerial complexities associated with marketing activities; and to ignore the importance, for producers and consumers, of a well-coordinated marketing system. We have found this image to be inadequate for purposes of developing marketing policies. Yet it remains as the conceptual cornerstone for policy formulation on agricultural marketing issues in most Latin American countries."[5]

Common misconceptions on price spreads in LDC's seem quite similar to those encountered in the U.S. in the recent past. Several writers have concluded that growing price spreads from the farm level to the retail level can be accounted for in an efficiency context by increasing proportions of services between the farmer and the consumer and by changes in the costs of these added non-farm services.[6] These same changes will likely continue to take place in the future. With rapid increases in petroleum prices and labor costs, these shifts have led to increases in the marketing margins for food products.

Experience in the lesser developed areas of the U.S. and other developed countries may be of considerable use in appraising alternatives in marketing in LDC's. The business organization of cooperatives is a marketing structure which is often recommended. Cooperatives are often thought of as an immediate panacea to all marketing ills in developing areas. The truth is that cooperatives often fail, as do other types of businesses.

Cooperatives require the same type of diligent management expertise and trustworthiness as any other viable business structure. The failures result from ill-planned ventures which are undercapitalized from the start, poor general management practices of the concern, and outright fraud. Fruit and vegetable cooperatives in Alabama are a case in point. Kern describes the poor results as follows: "The success of these markets was in part dependent upon anticipated production, which did not materialize. As farm adjustments occurred, resources were diverted to uses other than for vegetable production. This resulted in volumes of business insufficient to amortize market indebtedness at reasonable rates.

Other causes of failure subsequently reported from the several production areas included (1) lack of capable management, (2) the State assumed too large a part of the financing of the facility, and (3) the inability of markets to make full cash settlements to producers for their product at time of delivery. Another weakness was that producers failed to understand the cooperative basis, or were not willing to operate under such conditions."[7]

Many dollars have also been invested in cooperatives for low income producers. Poor management and fraud seem to be prevalent in these types of cooperatives the same as in other businesses. Farmers are often left disillusioned after they have lost their capital investment. Contacts with fisheries officials in several foreign countries have shown that similar poor results with cooperatives are often encountered in this industry. A large amount of feasibility research is in order to provide a sound basis for decision making on cooperatives as a prime market outlet for the small-scale producer. The producer does not have the ability on his own to assure that a proper study of feasibility is provided. At its best, the cooperative can be an excellent outlet for products as well as a buying entity for inputs used by small-scale producers. Caution and careful study of the local situation are necessary before starting such a venture.

Larzelere recognizes the possible advantages of cooperatives in aiding marketing in LDC's, but he qualifies the advantages by the statement that "Adequate education of potential farmer members is essential, especially with the realization that a cooperative must be programmed to perform some or all of the middlemen's functions better and more efficiently than is currently being done, and with the realization that some of the services, costs, and margins of present middlemen cannot be eliminated by a cooperative. Education in management skills to direct the cooperative is a necessity where cooperative formation is contemplated."[8] Knutson has stated two conditions necessary for successful cooperative development: (1) co-ops should have as their goal to equal but not exceed the market position of competitors and (2) to remain pure and family farmer oriented.[9]

The government itself can take a very direct part in the marketing of products. Governments may act as the sole buyer of products in some instances. In other situations the government may support special infrastructure, regulate, and operate special facilities to aid the movement of food products through the channels from producer to consumer. In some cases the government may not participate directly in the marketing, but may provide aid by subsidizing or furnishing facilities, furnishing credit to specified organizations, or providing grading and information services to an area. The government often aids cooperatives through some of the above means. Some of the difficulties in government price setting where the government is the sole purchaser of commodities are listed in the following sections.

In areas of low education levels of the populace, it may be necessary to have a strong central force to coordinate markets, to furnish facilities, to regulate activities and to set standards. The latter functions of the government may be necessary in any type of market system. The various activities must be coordinated from the producer to the consumer.

Infrastructure

The general economy's social infrastructure is a limiting factor on the marketing decisions for the business firm. Small-scale producers have the option of supporting political settings to further their aims through improved social infrastructure such as transportation, energy and communication utilities, port facilities and others which have spillovers to the aims of marketers. The commonly recognized marketing functions of storage, transportation, and financing depend in large part on these political decisions. The small-scale producers rarely have concentrated market power that can be directed towards reflecting their aim in the political sphere.

Major infrastructure such as highways, ports, water supplies, and other major investments cannot be financed solely for narrowly defined food crops in general. The food marketing system can complement other sectors of the economy in improving overall resource utilization. The most important objective of small-scale marketers is to be able to fit into the appropriate system of market structure and infrastructure to attain efficiency goals. Local decisions related to producers and assemblers must be coordinated in the total hierarchical structure in the manner required to reduce conflicts and contribute to the synergistic composition of the total economy.

THE ROLE OF THE MARKET AND PRODUCERS' DECISION MAKING

Agricultural markets in LDC's are believed to be of a more imperfect nature than in developed nations. The degree of severity in distortions is greater because of the distribution of resources leading to more incidences of market inefficiencies. Producers' interfacing with local markets can explain how decisions are formulated.

Information Flows

Availability of market information for producers can be

characterized as disorganized and unstructured. Markets, which are both dispersed and small in size in rural areas, can leave producers unable to formulate long-term marketing strategies. Without adequate information, producers are unable to maximize returns to farm enterprises. If information flows are disrupted, bargaining power for the producer is diminished.

An imbalance in marketing power leads to future inequities in credit and purchases of necessary inputs. This situation has been observed among tambak producers in Aceh, Indonesia.[10] Producers became entangled in interlocking credit schemes with the major buyers of their product. Producers then found themselves burdened by long-term marketing agreements to pay off loaned capital.

Oligopsonistic Buyer Behavior

In rural areas, dispersed, thin markets, especially for bulky products, can find buyers with enormous bargaining power. For example, in rural markets in Northwest Ghana, market days rotated among specified villages.[11] The major means of transportation were by truck which carried buyers to the markets. These trucks would be owned by wealthy traders in the local urban area. Ownership of the major means of transportation gave the buyers a market advantage. Market decisions were limited for producers because of a lack of access to alternative outlets for their commodity. These market intermediaries can foster price instability for producers.[12]

The producers' ability to counterbalance concentration in the buyer market is of major importance. In the United States, laws have been passed to allow an exception for producers of agricultural products to form organizations for negotiating and selling their products. This legal framework has given agricultural producers a method of improving their marketing performance, an institutional alternative not generally found in LDC's.

Lack of Entrepreneurship

The lack of growth in the numbers of entrepreneurs in LDC's is seen as a deterrent to increased market performance. Market systems in LDC's are made up of sellers who have low levels of technology available to them. They sell small lots making a subsistence earning by taking a small margin on each unit sold. It is not uncommon in West African markets to see several middlemen performing services that could be performed by one person. The cultural arrangement is such that each person respects the activity performed by the person at the next stage of the marketing system. Entrepreneurs who provide risk capital for these market functions are innovative in the adoption of new technologies.

Small-scale agricultural producers find it difficult to assume additional risks associated with being an entrepreneur. A limited resource base prevents adoption of appropriate tech-

111

nology that could reduce costs and enhance product sales. The lack of entrepreneurs often means markets will remain stagnant without an enrichment of ideas to improve technical efficiency.[13] This effect is felt by producers who have little incentive to change their marketing practices.

Government Price Controls and Stabilization Policies

Government's role in market operations has been seen in varying degrees of control. In the extreme, markets can be completely unregulated or government buyers may be the only purchaser. Many examples abound of government marketing firms that are licensed as the only exporter. Small-scale producers can receive low prices compared to the export market because the government needs revenues to support its programs. Government buyers cause lower prices to be offered to producers causing disincentives for adoption of technology and a decreased commercial supply response.[14] Excessive government control of the market functions can distort price signals for allocative efficiency.[15] Government's policy to stabilize prices and producers' incomes could lead to greater instability in the marketing system.

In extreme examples, black markets and corruption accompany disequilibrium prices which are set too low by the government. Many examples exist in Central and West Africa in which maximum prices of products led to the abandonment of coffee and other cash crops due to excessive price controls. Corruption of government officials accompanied by channeling of crops through the black market is also common in such circumstances.

DECISIONS RELATED TO THE NATURE OF THE PRODUCT

Marketing decisions are limited in many ways by the nature of the potential product and its place in the overall pattern of resource use. A tremendous number of variables is involved, allowing spillovers of effects of one variable to another in the market. Marketing policies must recognize the great number of tie-ins and the effects of adjustments from one element in the market on other elements. Some of the features and requirements of products as related to marketing decisions are discussed below.

The Resource Base

Small-scale entrepreneurs may be somewhat limited in their decision making by the nature of the resource base in their area. This limitation may be complicated by rigid land tenure arrangements in the locality. Resource adaptability is the first constraining criterion on production which is transferred to marketing decisions. Soil and/or water characteristics may be so enterprise-specific in their adaptability that few choices are available concerning what to produce. Usually several alternatives are available which necessitate a careful appraisal of opportunity cost in selecting enterprises. The narrower the list of alternatives of production, the fewer will be the number of choices.

Special efforts must be made to find the proper enterprise to fit local resource bases in cases of unadaptable resources. Experimental programs have been carried out in El Salvador at a special station for brackish water aquaculture to further exploit an aquaculture base that has been underdeveloped heretofore.[16] Extensive brackish water aquaculture projects are also being carried out in other parts of the world, especially in the growth of certain shrimp products.[17]

In El Salvador, David Hughes reports an extreme case of adaptability of a resource to added production by use of salt production ponds for production of a crop of tilapia for marketing during the six month off-season of the salt evaporation process. As much as 1,000 kilos per hectare of fish were grown for market during the slack period of evaporation. The fingerlings were put through an acclimatization process to the saline waters in a gradual manner preventing shock so that fresh water fish could develop as if they were saltwater species. The tilapia harvest was essentially a "free good" above the normal salt production of the enterprise.[18]

The complementary production processes should be exploited when the opportunity presents itself. Irrigation ponds or other special purpose resources often can be used in growing fish or other polyculture harvests for sale.[19]

In contrast, Jamaican land resources were not found adaptable to pond fish culture. Much of the land was of a gravelly nature and would not hold water. Low water-holding ability, few sizable streams, the steep and rocky topography and millions of natural sinkholes in the limestone base rendered a large part of the area useless for pond fish culture.[20] Porous gravelly lateritic soils and sandy soils also rendered many of the West African areas unsuitable to pond fish culture.[21]

Supply Stability Decisions

Several types of opportunities exist for supply instability in the provision of food products to the market. Cobweb-type instability may exist when the annual supply of crops is based on prices in previous time periods. This condition develops when the dependent-independent nature of price and quantity reverses itself in alternate time periods. The fluctuations can be quite serious when individual decisions are made on production and in the aggregate lead to the detriment of the entire group. Many LDC's are in tropical areas which may or may not have lags of a seasonal nature and cobweb-type supply and price fluctuations. Efficient marketing on a large scale suggests efforts to smooth out fluctuations over time, especially with respect to commodities not likely to be storable. Fruits and vegetables available from annual croppings

113

show need of a central coordination to prevent serious disruptions.

Seasonal fluctuations also have adverse impacts on the market in some instances. Optimum planning of production and storage can often lead to great benefits to sellers of products. Some low-income buyers, however, can afford the price of certain commodities only when their prices fall to unusual levels.[22] These consumers can be eliminated from buying by smoothing out seasonal variations and cobweb variations. In the case of aquaculture, fish products can be timed by the seller to bring about a counter-seasonal supply of products to take advantage of natural shortages from seasonal changes.

Export Markets versus Local Markets

The choice of market outlets, local and export, may not be a decision relevant to the small-scale producer, but in some cases he will have the choice of options. A comparison of requirements in the two types of markets might be in order. Several criteria for comparison between local and export markets might be: 1) assembly facilities, 2) grading, 3) packaging, 4) uniform timing, 5) financing and credit, 6) a central organization, and 7) disbursal of dependable information.

In the export market, assembly facilities are a necessity to provide adequate quantity concentration for shipment on a commercial scale. This market also requires standard grading and packaging since buying and selling on a large scale can be done only on an impersonal basis in which good product descriptions can be made. Dependable timing is also of great concern when operating in an export market. Outside financing and credit may be a necessity, when dealing with an export market, to bring dependability to supplies. This latter requirement makes the selection of a central organization important. The coordinating mechanism might be a government or a private entity such as a cooperative. The last major requirement is that an adequate information scheme be used in order to promote the export product.

In contrast to these rather strict requirements in the export market, local sales may require no assembly facilities or only scant facilities. It may be possible to sell without grading, by sight purchases; and qualities which would not be acceptable for export could be perfectly acceptable to local customers. Only rudimentary packaging is required in the local markets where customers often furnish their own conveyances. Timing is not as critical for local markets since customers can observe availability. Financing and credit are generally not a concern in local markets. No central organization is required and information is available on the spot.

Capital Intensive versus Labor Intensive Products

The determination of types of products to produce in an area is somewhat limited by factor availability and governmental attitudes of factor use, especially in areas of high unemployment rates for labor. Labor-intensive production and harvest methods may be required in some areas in an attempt to lessen unemployment. Disproportionate labor quantities may be required in planting and cultivating over harvesting or vice versa.

The marketing also may require a large or small proportion of the total efforts between the producer and the consumer. Careful selection of types of crops may stagger peak labor requirements to smooth out labor demands to better utilize labor supplies. A systematized selection of products may therefore increase efficiency and add to stability of the community through more productivity and a more nearly full utilization of resources.

Traditional or New Products

Decisions on traditional versus new or innovative products are difficult for the small-scale producer because of the considerable risks with respect to his knowledge of the market. The small-scale producer will often be ignorant of product options open to him within a production context and he will usually have even less knowledge of the market feasibility of innovative products. Some outside agency usually must guide potential producers through the steps and create an information base on the acceptability of new products. An extensive promotion may be required which is beyond small-scale marketers, even on a group project basis.

Consumer demonstrations may be required in creating a demand for a product. In some areas, superstitions and taboos may form serious barriers to production and marketing of nontraditional food items in local or regional economies.

Incomes of Potential Consumers

Although the small-scale marketer may be somewhat isolated from the overt decision making on products concerning income levels of potential customers, this variable and its changes can be of considerable importance in marketing successes. Estimates of income elasticity from advisory authorities would be useful for planning, but guidance on the nature of the populace with respect to income responses in terms of normal goods (those whose demand increases, other things equal, when income increases) and inferior goods (those whose demand decreases when income increases) would aid greatly in individual decision making. In periods of falling income the provision of inferior goods can benefit the seller and consumers. "Inferior" in the economic sense does not imply low quality in a nutrition sense and should not suggest low esteem of the product.

Segmentation of the market may be necessary to supply different types of products to the high-income populace and to low-income and intermediate-income groups to maximize profits of producers and most efficiently use market facilities. This action may entail production of goods for low-income buyers at local markets and different goods for high-income export markets such as those discussed in the preceding section. The small-scale producer in LDC's will generally need information from an outside agency to facilitate his production and marketing decision making on these matters. The level of expertise in marketing by producers needs to be higher than under traditional systems.

Perishability-Storability Considerations

The nature of the product is of extreme importance in planning the production-harvesting-marketing strategy at the local level. Perishable products must be either sold on the spot for local use or stored or transported with adequate cooling or other necessary care for future use. The availability of adequate storage and disposal infrastructure and knowledge about care of products is essential to the latter alternative. Long-lasting product image adversity has arisen in some instances due to abuses of product handling and storage methods.

Aside from cooling perishable commodities, other common methods of preservation are smoking, drying, salting, and canning. In dealing with fisheries products, some of these methods which are effective and adequate when followed in a technically correct manner, have been abused by producers waiting until product quality has deteriorated badly, then smoking the fish, putting them on ice, or salting them. The potential customer acculturated in such a setting entertains the question of why anyone would smoke the fish, put them on ice, or salt them if they had not deteriorated badly. The seller is in the grips of past mistakes. This adverse image from abuses precludes the entrepreneur who uses correct technical methods from being rewarded for the necessary care required to turn out a nutritional product in the market.

A similar product perception problem has arisen in which a seller of fish would spray the fish with flyspray to keep insects away. A serious potential for poisoning of the consumer existed, leading to consumer preference of fish in the market with plenty of flies swarming around them to indicate that the fish were safe for eating. There must be many parallels to these examples in the handling of other perishable products in the market.

The major way to break down this market inefficiency is through education of the marketer on methods, education of the consumer on quality judgment, and inspection for enforcement of quality maintenance by responsible officials. The scale of the marketing outlet will determine if the latter alternative is feasible. It is much easier to create and maintain a good image from the beginning than to restructure and remove the tarnish of a bad image of the commodity.

Adoption of proper technology for food preservation and

marketing is a complementary operation to the education of marketers and consumers. Any reduction of wastes poses possibilities for improving the efficiency of the system and passing on cost reductions to the ultimate consumer.

PRODUCER'S PERCEIVED RISKS

Agricultural producers face innumerable risks in production and marketing of their commodities. These risks become increasingly severe as farmers approach subsistence levels and affect their willingness to accept new technology.[23] Understanding levels of acceptable risk preference of producers can give insight into what projects are likely to succeed for peasant producers in improving their marketing.

Research on the effect of the high yielding varieties has indicated that availability of factor inputs has led to new problems in factor utilization, marketing policies, and employment and equity disparities.[24] Timeliness of factor supplies is crucial in any agricultural production system, especially when favorable weather may prevail only for a short duration. Factor markets for small-scale agriculture may be either in the private, public, or a combination of the two sectors. Government policies have generally focused on public sector control of important raw resources necessary for the "high yielding" technology. Inadequate or centrally-controlled infrastructure can lead to the breakdown in distribution of the needed resources. For these reasons, producers could be wary of adopting technology to enhance production and marketing.

On the output side, market decisions by producers influenced by factor availability can tailor marketing strategies which do not necessarily promote efficient marketing. Distortions in the marketplace for outputs cause inefficiencies that affect all aspects of a producer's decision-making process. For example, Sullivan et al. found that in Tanzania livestock producers experiencing low government controlled prices retained animals in their herds beyond the physically optimum age for slaughter.[25]

Agricultural development that has as its aim to increase production without the proper economic incentives for prices of products will find technology adoption lagging.[26] Public sector control of market prices, even to the point of having a sole buyer, creates risks to producers of not knowing what government policy will be in the next time period. This risk is especially true if government administration is unstable. All too often, government policies have been to the benefit of other sectors of the economy at the expense of the agricultural sector.

Often overlooked at the farm level are the excessive amounts of corruption that producers experience in obtaining necessary supplies, transportation, or purchases that require payments to government officials. If market prices are kept

117

low, the added "kick-backs" to officials could make costs high enough to detract from successful involvement of producers in a project. The transactions costs for participation can make involvement unattractive.

Bottomley has discussed the credit markets as having characteristics of being skewed against small-scale producers.[27] Farmers' lack of credit can cause them to sell crops at harvest rather than put the crops in storage. Producers, unable to hold crops in storage, sell when prices are normally at their seasonal lows. Producers' risks increase when storage facilities are inadequate, causing more deterioration than is normal. A project conducted by a private volunteer organization in Northwest Ghana gave support to build village storage facilities so farmers could store crops and obtain money from the village union to pay their debts while still keeping ownership of their crops. Crops could then be sold when prices were higher during the dry season. This action also allowed producers more bargaining power in dealing with food buyers attending their markets.

INFLUENCES ON PRODUCER DECISION MAKING

Figure 8.1 illustrates the linkages influencing a producer's marketing decision. The model is based on the Robinson, et al. framework for buyer-consumer behavior theory.[28] On the left side of the figure are conditions embodied in the environment which impact on the producer as well as the institutional arrangements in the market system. Broad general conditions, macroeconomic conditions, microeconomic conditions and countervailing power comprise the environment. General conditions influence macroeconomic and microeconomic policies as well as other producers selling in the market place and resource utilization practices. Countervailing power means the ability of producers to offset power of buyers in market transactions.

As discussed earlier, characteristics of market organization also directly influence the producer's decision-making process. Factors such as number of sellers and buyers, types and quality of infrastructure, types of buyers and the completeness of information all make up the organizational constraints. These factors directly affect the day-to-day decisions on how producers will market their commodities. The process of improving traditional systems is influenced by emotional attitudes of a producer. The degree of allegiance to tradition will slow the diffusion of adoption of marketing technology. Culture and religion also have a role to play in this process. The producer's psychological make-up influences seller behavior and the learning process of the individual. Finally, economic consideration about household constraints are important. The need for security in family income, subsistence and nutrition, and perceived risks of obtaining these economic goals all influence the producer's decision-making process.

Figure 8.1. Diagram Showing the Interactive Relationships Among Variables Influencing the Marketing Decisions of Producers

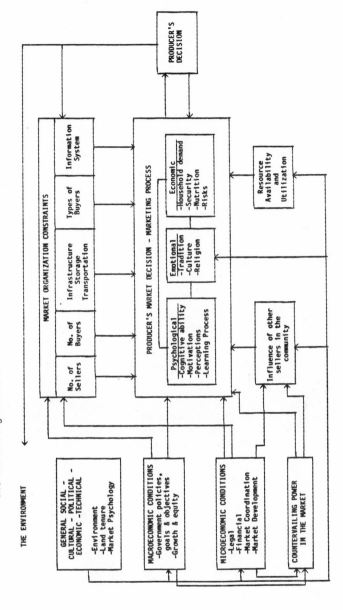

119

The final decision derived by a producer is a combination of the internal factors interacting with the environment and market organization constraints. The feedback mechanism is the producer's selling decision on the environment over time.

POLICY CONSIDERATIONS

The design of appropriate government policies for improving technology adoption is a concern of all development planners. Project planners need to be aware of how producers set their marketing strategies. Economic incentives, both perceived and real, influence producers' decision-making processes. When incentives become distorted and unclear, producers have difficulty in setting consistent market practices. Consequently, adoption of new technology has inherent increased economic risks.

Economic development projects need to be attuned to facilitating producers' marketing decisions. The need for appropriate institutions should have a high priority. Facilitating functions of the market, like grades and standards, quality, and market information, improve the operating efficiency of a marketing system. Producers benefit from these kinds of market improvements. Over time, they can reduce risks inherent in a marketing system.

The mechanisms by which producers' risks are minimized have a synergistic effect on the whole technology package for the producer. Institutions, which improve producers' market power, lead to an overall improvement in performance. Market coordination and bargaining associations propel producers to discover a new set of activities and knowledge. The spin-off effects enhance development of human capital.

Certain types of producer organizations are heralded as remedies for problems associated with small-scale producers. Cooperatives are a typical example. Success and failure of these recommendations lie in the void of inexactness of social science. A model for replication is difficult to reproduce under varying cultural, economic, social, and political systems. Each example is on a case by case basis with only broad generalities likely to emanate.

Any complex system generates secondary welfare considerations. Producers' decisions have to be regarded in light of equity considerations. The stream of benefits needs to be measured in light of its economic distribution. Enhancement of one set of producers could be at the expense of another set which could be smaller or larger in number. There is a need for measurement and identification of these benefits to each group of participants.

REFERENCES

[1]Christos Kamenidis, "The Anti-Middlemen Attitude of

Farmers in Greece: Causes, Repercussions, and Solutions," in The Rural Challenge, eds. Margot A. Bellamy and Bruce L. Greenshields, eds. (Hampshire, England: Gower, 1981); Luis F. Herrman, Openers Remarks to "The Anti-Middlemen Attitude of Farmers in Greece: Courses, Repercussions, and Solutions," In the Rural Challenge, eds. Margot A. Bellamy and Bruce L. Greenshields (Hampshire, England: Gower, 1981).

[2]Ibid.

[3]Ibid.

[4]Henry E. Larzelere, Opener's Remarks to, "The Anti-Middlemen Attitude of Farmers in Greece: Causes, Repercussions, and Solutions," in The Rural Challenge eds. Margot A. Bellamy and Bruce L. Greenshields, (Hampshire, England: Gower, 1981).

[5]Kelly M. Harrison, "Public Policies and the Development of Effective Marketing Systems" (Paper presented to the Agricultural Policy Seminar sponsored by Inter-American Development Bank, Washington, D.C., March 16-21, 1975), p. 1.

[6]Jerome W. Hammond, Willis E. Anthony, and Martin K. Christiansen, "Why the Growing Farm-Retail Price Spread?" in Agricultural Policy in an Affluent Society, eds. Vernon W. Ruttan, Arley D. Waldo, and James P. Houck (New York: Norton, 1969).

[7]E. E. Kern, "Marketing Truck Crops in Alabama," Bulletin 314 (Auburn, Alabama: Auburn, University, March, 1959), p. 6.

[8]Henry E. Larzelere, "The Anti-Middlemen Attitude of Farmers in Greece."

[9]R. Knutson, "Cooperative Strategies in Imperfectly Competitive Market Structures - A Policy Perspective," American Journal of Agricultural Economics 56 (1974):904-12.

[10]G. M. Sullivan, "A Marketing Study for Milkfish and Shrimp from Brackish Water Tambaks in Aceh Province, Sumatra, Indonesia," (Auburn, Alabama: International Center for Aquaculture, Auburn University, May, 1981).

[11]G. M. Sullivan, "An Economic Analysis of the Impact of Marketing Cooperatives on Rural Development" Unpublished Master's paper, Texas A&M University, November, 1974).

[12]J. Bieri, "Market Intermediaries and Price Instability: Some Welfare Implications," American Journal of Agricultural Economics 56 (1974)

[13]F. M. Scherer, Industrial Market Structure and Economic Performance (Chicago: Rand McNally College Publishing Co, 1980), p. 632.

[14]G. M. Sullivan and D. E. Farris. "The Role of Government Policies in Agricultural Development: The Case of the Livestock Industry in Tanzania." Zeitschrift fur Auslandische Landwirtschaft 21 (1982):52-61.

[15]T. W. Schultz, Distortions of Agricultural Incentives (Bloomington: Indiana University Press, 1978), p. 350.

[16]Donald R. Street, "The Socio-Economic Impact of Fisheries Programs in El Salvador" (Auburn, Alabama: International Center for Aquaculture, Auburn University, February 1978).

[17]Richard Pretto-Malca, "Aprovechamiento de Las Aguas y Excretas de la Explotacion Porcina Para el Cultivo de Peces en Panama," Revista Latino Americana de Aquicultura, 3 (1980):29-33.

[18]Street, "The Socio-Economic Impact of Fisheries Programs in El Salvador."

[19]Street, "An Economic Assessment of Jamaica's Fish Culture" (International Center for Aquaculture, Auburn University, August, 1978).

[20]Street, "An Economic Assessment of Jamaica's Fish Culture."

[21]John H. Grover, Donald R. Street, and Paul D. Starr, "Review of Aquaculture Development Activities in Central and West Africa," (International Center for Aquaculture, Auburn University, November 1980.)

[22]Donald R. Street and Gregory M. Sullivan, "Equity Considerations for Fishery Market Technology in Developing Countries: Aquaculture Alternatives" (Department of Economics, Auburn University, 1982).

[23]J. L. Dillion and P. L. Scandizzo, "Risk Attitudes of Subsistence Farmers in Northern Brazil: A Sampling Approach," American Journal of Agricultural Economics 60 (1978):425-435.

[24]Walter P. Falcon, "The Green Revolution: Generations of Problems." American Journal of Agricultural Economics 52 (1970):698.

[25]G. M. Sullivan and D. E. Farris. "Survey of Traditional Livestock Industry." Tanzanian Livestock-Meat Subsector vol. II (College Station: Texas A&M University, October, 1976).

[26]Ibid.

[27]A. Bottomley, "Monopoly Profit as a Determinant of Interest Rates in Underdeveloped Rural Areas," Oxford Economic Papers 16 (1964):431-437.

[28]Patrick J. Robinson, Charles W. Faris, and Yoram Wind, Industrial Buying and Creative Marketing (Boston: Allyn and Bacon, Inc., 1967).

9

Sociocultural Constraints on the Transfer and Adoption of Agricultural Technologies in Low Income Countries

William L. Flinn
Frederick H. Buttel

INTRODUCTION

Perhaps no social science literature has changed so rapidly and dramatically over the past 20 years as has the "development" literature. In the early 1960's there were a shared confidence and a commonality of purpose among economists, sociologists, anthropologists, and political scientists concerned with understanding and enhancing Third World development. The early 1980's find these literatures increasingly fragmented along disciplinary lines and theoretical postures.

"Induced innovation"[1], "dualistic development"[2], "dependency"[3], "modes of production"[4], "moral economy of the peasant"[5], the "uncaptured peasantry"[6] and the "rational peasant"[7] are just a handful of the major notions that now constitute specialized literatures, the authors of which tend to be increasingly isolated from one another. This situation has been confounded, moreover, by the rapid succession of slogans and programs (e.g., integrated rural development, basic needs, appropriate technology, farming systems research) promulgated by development agencies in First and Third World countries.

The research traditions mentioned earlier have had uneven impacts on agricultural development practice. In fact, development practice has tended to be dominated by a relatively small number of approaches, basically being limited to micro-level perspectives utilizing nonthreatening concepts. This is not entirely the fault of development professionals, since, for example, "modes of production" analysts have exhibited little inclination to think through the applied implications of their work. Nevertheless, development practice is now based on a rather circumscribed slice of the development theory pie--a circumstance which we argue below has led to an incomplete understanding of the present and future constraints on technology transfer and adoption.

The principal point in this chapter is that social structures internal and external to the Third World are experiencing rapid changes, with implications that have only incompletely been incorporated into the practice of technology transfer and agricultural development. The major purposes of the paper will therefore be to identify the key structural transformations that directly and indirectly affect agricultural development and to take some steps toward specifying their implications for agricultural development practice. It should be noted that our comments will be rooted most strongly in the Latin American experience, the region with which we are most familiar. Nevertheless, we would suggest that these observations have a more general applicability even though African and Asian specialists will no doubt take exception to some of the specifics of Latin American agricultural development that we advance as pertaining to the Third World as a whole.

THE BASIC FORCE: THE EVOLUTION OF THE
INTERNATIONAL DIVISION OF LABOR

It can now be seen in retrospect that the ferment in development theory during the 1970's largely centered around whether Third World development could be understood in terms of the socioeconomic forces that shaped the transformations of the presently advanced industrial societies of Western Europe and North America.[8] We think the weight of the evidence is in contradiction with the early influential development theories that essentially posited the replicability of the Western path of economic development in the Third World.[9]

It has become increasingly apparent that there are a number of factors that render the Western experience essentially nonreplicable in most areas of the contemporary Third World.[10] First of all, the present developed economies generally did not face the problem of colonial pillage; contemporary Third World countries likewise cannot take advantage of abundant capital and the extraction of raw materials from other continents or cultures that historically assisted the development of "center" or advanced industrial societies.

Second, the present underdeveloped nations are generally plagued by overpopulation and do not have the possibility of reducing their populations through international migration, as was the case with much of Western Europe during the 1700's and 1800's. Third, the contemporary advanced industrial nations did not face protectionism and trade monopolies with which the underdeveloped countries must now contend. Fourth, center countries did not have to attempt to foster an individual and agricultural take-off from a position of technological inferiority relative to already developed countries, as is the case with contemporary countries on the periphery.

Because of the advanced state of current industrial technology, especially its labor-saving character, "hand-me-down" technology that goes to the Third World has little labor-absorbing capacity. To be competitive on international markets, Third World countries must purchase capital-intensive industrial technologies that employ relatively few persons and hence prevent the emergence of a mass industrial working class as has occurred in the West.

Because of the lack of a mass industrial working class, present-day underdeveloped nations face severe barriers to economic growth expansion of the internal market. Instead, market expansion must be heavily (although, of course, by no means entirely) oriented toward external, export markets, and/or toward a handful of affluent families, and deprives the peripheral economy of backward and forward linkages and multipliers of economic activity.

The structure of the international division of labor is characterized by three major factors, one of which has long characterized the world economy but two of which are largely unprecedented. The first characteristic of the international division of labor is dominance and subordination among nations. These disparities are not confined only to statistical differences in living standards. What is particularly crucial is the technological superiority of First World nations, their greater levels of capital, and their dominance over the patterns and conditions of world trade. The distinction between career or advanced industrial nations on one hand, and developing or peripheral nations on the other, thus reflects the various aspects of the nonreplicability of the Western path of industrial development that we spoke of earlier.

The second and third characteristics of the international division of labor are the unprecedented degrees of market penetration in Third World economies and of capital mobility across national boundaries. As factor and product markets have become increasingly international in scope, peripheral economies have become subject to the promises and pitfalls of comparative advantage. Moreover, economic enterprises are increasingly willing and able to change locations in search of more profitable investment opportunities. The result is that international economic relations increasingly are market relations on a world scale, increasingly subjecting Third World economies to the vagaries of market competition.

Technological inferiority in conjunction with increasing international economic interdependence has heavily shaped the nature of Third World industrialization and ultimately of economic development and agricultural policy. For industrial development to proceed in a milieu of technological inferiority and international economic interdependence, Third World countries must import capital goods from the advanced industrial societies. The infrastructure necessary to support industrialization requires economic policies that ensure sufficient foreign exchange for debt service. This will typically dictate economic policies which subsidize the export sectors (both industrial and agricultural) over sectors primarily oriented toward the internal market.

In order for export-oriented economic policies to be successful under conditions of international economic interdependence, Third World industry must be economically competitive. As noted earlier, this imperative dictates the utilization of relatively advanced, capital-intensive technologies that absorb relatively little labor but which constrain growth of the internal market. Nevertheless, labor costs are a critical component of industrial production costs, since the firms of advanced industrial states tend to confine the deployment of the very latest and most efficient technologies to their home countries.[11] Thus, peripheral countries must compete internationally primarily on the basis of low wage scales.

The governments of developing countries tend to orient economic policy toward minimizing labor costs. Policies supporting low wage scales are varied, but include public subsidies of workers' consumption costs (e.g., transportation and other services) and, most importantly for the purposes of this paper, policies that ensure low food prices.

The tendency for Third World governments to enact policies which result in low food prices has been widely recognized by conservative[12], liberal[13] and radical[14] observers. Cheap food policies are pursued in direct fashion (e.g., manipulation of currency exchange rates, importing of feedstuffs) and indirectly through ensuring cheap agricultural labor. A major example of the latter aspect of food policy is the initiation of integrated rural development programs aimed at retaining peasants in the countryside, but at a sufficiently low level of subsistence so that they are available as a cheap labor force for the estate sector.[15]

A final feature of the international division of labor that is important to emphasize is the growing socioeconomic differentiation among what are variously described as low-income, developing, Third World, or peripheral countries. There has been a strong tendency within otherwise disparate development theories to view late-industrializing countries as a relatively homogeneous group characterized by similar dynamics. This assumption is no doubt warranted in a general sense, and we will indeed later succumb to making certain generalizations about Third World nations as a vehicle for simplifying the discussion. Nevertheless, there has been a growing recognition that international economic interdependence and the uneven impact of the forces of comparative advantage have led to substantial and growing disparities among low-income countries.[16]

Historically, the inability of traditional neoclassical economic development theories to cope with the nonreplicability of the Western path of development led to the emergence of a "dependency" perspective to explain the apparent stagnation of the peripheral economies. Baran argued in a preliminary discussion of what is now known as dependency theory (the first Western scholar to do so), that the development of the advanced industrial countries was and continues to be preconditioned on the underdevelopment of the periphery.[17] That is, the world economy was seen to function so as to extract surplus from periphery to center. In Baran's view, the peripheral economy is essentially a mechanism for extracting surplus from the Third World, leading to underdevelopment and stagnation.

The notion that dependency is inherently accompanied by economic stagnation became the hallmark of mainstream dependency theory.[18] In the terms of dependency theory, the problem of Third World food production was essentially reduced to one of economic stagnation caused by external dependency. Economic stagnation implied capital scarcity and extraction of value from agriculture to the primate city, and from the peripheral primate city to the center (through debt repayment, royalties or payments for capital goods, imports to luxury goods, etc.). The limit to food production was therefore the external dependency and general economic stagnation set forth by the conditions of subordinate participation in the world economy.

This wholly stagnant image of Third World agriculture and industry cannot be sustained from the available data.[19] It has become increasingly apparent that a set of Third World countries--especially Taiwan, South Korea, Brazil, Argentina, and Chile--have tended to exhibit relatively rapid rates of economic and industrial growth. For most of these countries, the proportion of GNP accounted for by industry is in excess of the share accounted for by agriculture. These countries are well on their way to becoming major industrial powers--with some, such as Brazil, having significant capital goods production capability--and belie classical dependency theory's image of being largely agrarian economies with small agricultural or mineral enclaves producing solely or primarily for the export market.

Basically, one can identify four major categories of peripheral countries. The first, the rapid-growth, semi-industrialized countries which were mentioned above, are often referred to as "semiperipheries," following Wallerstein.[20] These countries provide major markets for capital goods and attractive investment opportunities for multinational manufacturers of luxury consumption commodities. Also, as noted earlier, the semiperipheries tend to have substantial capital goods production industries owned, at least in part, by home country firms or by the state.[21]

The second set of Third World countries is the oil-exporting countries such as Saudi Arabia, Kuwait, Nigeria, and Venezuela. While the oil-exporters are themselves diverse, they share a common pattern of receiving massive infusions of capital from oil importing states and having relatively low energy prices and favorable trade balances. Their development problems are heavily oriented around rational utilization of oil revenues to achieve rapid, but balanced growth without inflation.

The third category of peripheral countries is a large group of export-oriented agricultural and mineral economies

129

such as Columbia and Thailand. These countries do not have a great deal of industrialization and are usually dependent upon the export of one or a few key raw materials commodities.

The final category of countries might be referred to as "lumpenperipheries", since these are countries which are still dominated to a high degree by merchant oligarchies, lack major exportable raw materials, have little or no industrialization, are typically overpopulated relative to food production capability, and are characterized by highly stagnant, largely agricultural economies. Haiti is the ideal-typical example of a lumpenperiphery which faces overpopulation, lack of resources, grinding poverty, and widespread hunger and malnutrition. These categories are not meant to be mutually exhaustive or exclusive because Third World diversity defies such convenient classifications. The four categories nonetheless impose some order on the diversity of low-income countries which has yet to be systematically incorporated into prevailing development theories.

THE TRANSFORMATION OF THIRD WORLD SOCIAL STRUCTURES: SOCIOECONOMIC CONSTRAINTS ON TECHNOLOGICAL CHANGE AND AGRICULTURAL DEVELOPMENT

The foregoing has emphasized the role of the international division of labor in conditioning structural change and economic policy in Third World countries. We do not want to take the extreme view that peripheral countries' social structures and economic development policies are wholly determined or induced by external relations in the world economy.[22] Nevertheless, international economic interrelations provide a key context for the ways in which internal socioeconomic forces impinge upon structural change and the organization of food production.

The present section consists of a general overview of some of the major ongoing changes in industrial and agricultural organization that have major implications for the transfer and adoption of agricultural technology. The admittedly overgeneralized nature of these observations should be kept in mind, as we alluded to earlier. The purpose of making these observations is not to provide ironclad generalizations about all Third World countries, but rather to provide insights into the nature of developmental tendencies and their implications for programs of technology transfer and adoption.

From Hacienda to Plantation
Less than 15 years ago, development theorists of virtually all stripes agreed upon the fundamental importance of encouraging land reform in the developing world. Dorner[23] and more recently Berry and Cline[24] made compelling economic--and, to a lesser extent, sociopolitical or moral--causes for land reform, emphasizing the inefficiency of feudal, semi-feudal, or precap-

130

italist estates relative to small peasant producers. It is ironic that the most convincing empirical case for this argument, by Berry and Cline,[25] was made on the basis of relatively old data at the very time that traditional estates were being transformed into plantations subject to the discipline of competitive factor and product markets.[26]

We do not wish to posit the inherent technical superiority of modern estates over small peasant producers; this conclusion would be no more warranted than the narrow economic case for land reform that we believe can no longer be sustained (at least universally). Nevertheless, as global economic interdependence has tended to transform the precapitalist estate into a more efficient agricultural production organization, the context of technology transfer and adoption has changed considerably.

The rationalization of precapitalist estates means that large farmers are more receptive than ever to adoption of new technology. Moreover, these large farmers are becoming increasingly organized to influence the research priorities of agricultural research institutions. Finally, these increasingly efficient large farms are becoming preeminent as the locus of capital accumulation in the agricultural sector; this not only ensures perpetuation of their political power but also reinforces their position as the major market for imported and domestically produced agricultural inputs. Thus, both political and market factors suggest that the estate sector will become increasingly important as clients of agricultural research institutions and as customers for marketers of agricultural inputs.

Semiproletarianization and Re-peasantization

Much of the literature critical of the Green Revolution was based on the observation that Green Revolution technologies were large farm- or landlord-biased and that deployment of these technologies had led to the demise of small peasant farmers.[28] Subsequent research has demonstrated that while these generalizations tended to hold in many Third World contexts, the socioeconomic impacts of Green Revolution technologies were strongly dependent on the institutional arrangements within which they were deployed. More important for our purposes, however, is the need to examine more closely the notion that persistence of agricultural enterprises in the wake of adoption of Green Revolution technologies has been inversely related to farm size.

Although there are exceptions, there has been a general pattern in Third World agriculture for the state sector to be functionally related to the subsistence or small-holder sector through members of small-holder families working for wages (either cash or, decreasingly, in-kind) on neighboring large farms.[28]

131

The pattern is frequently referred to as "functional dualism" in an implicit critique of theories of dualistic development which tend to portray the "modern" (estate) and "traditional" (subsistence or peasant) sectors as being economically if not spatially separate.

The tendency toward functional dualism can be explained by the high labor demand (albeit often only seasonal) of the estate sector and the imperative faced by these large farmers to produce on an economically competitive basis; labor costs can be reduced by perpetuation of the subsistence sector which partially assumes the cost of maintaining the labor force. Functional dualism between the modern and traditional sectors thus enables large farmers to pay their hired workers a wage which is lower than the cost of maintaining the worker, since part of the subsistence costs of the labor force is provided by unpaid family workers who labor on small subsistence plots.

To return to the issue raised at the outset of this section, there has been growing evidence that the structural changes accompanying the Green Revolution can be more accurately viewed as a semiproletarianization process rather than a more straightforward proletarianization process. By semiproletarianization[29], we mean the twin processes of peasants' loss of access to sufficient land to support a family and of these marginalized peasant families having to rely increasingly on wages to supplement farm income and the production of subsistence crops.[30] To be sure, the expansion of the Green Revolution did displace many peasants, with large numbers of these peasants migrating to cities.[31] However, the Green Revolution appears to have been associated more with the decline of "middle peasants" than with the disappearance of marginalized semiproletarians, with many former middle peasant households being reduced to smallholders on tiny plots.

Subsequent research has demonstrated that while these generalizations about functional dualism tended to hold in many Third World contexts, the socioeconomic impacts of Green Revolution technologies were strongly dependent on the institutional arrangements within which they were deployed. More important for our purposes, however, is the need to examine more closely the notion that persistence of agricultural enterprises in the wake of adoption of Green Revolution technologies has been inversely related to farm size.

Peek[32] has reported data on 11 Latin American countries which reveal a strong trend in the direction of "re-peasantization"--that is, the reinforcement of peasant numbers on small, "subfamily" plots.[33]

For example, in Brazil, farms under 10 hectares in size increased from 34 to 51 percent of total farms from 1950 to 1970. During this same time period, the average size of

Brazilian farms under 10 hectares decreased from 4.25 to 3.61 hectares. Most Latin American countries exhibited comparable trends, with the tendency toward re-peasantization being strongest in those countries (Brazil, Ecuador, and Colombia) with the greatest degrees of land concentration.

The data reported by Peek suggest that re-peasantization has been closely connected with the growth of the modern estate sector in agriculture.[34] However, it should be kept in mind that agricultural wage employment in Latin America increased by only 0.5 percent per year from 1950 to 1970, accounting for 24 percent of the growth of the total agricultural labor force. Since farms less than five hectares accounted for 76 percent of the growth of the Latin American agricultural force during this same time period, it is apparent that the subsistence sector is not only a source of cheap wage labor but also a refuge for the growing and increasingly underemployed rural population facing inadequate wage employment opportunities.

The trends toward semiproletarianization and re-peasantization have several implications for technology transfer and adoption. First, given the shrinking farms operated by peasants, this peasantry, despite its growth in numbers, is losing its status as commodity producers (as opposed to producers of foodstuffs for home consumption). Shrinking plots diminish the possibility of peasants producing marketable surpluses, making these near-landless peasants increasingly unlikely to assume the risks of adopting new agricultural inputs. Second, the growing numbers of peasants relegated to more and more minute and eroded plots are unlikely to be able to capture the benefits of increased agricultural productivity even if they do utilize improved technologies. Research in Latin America has indicated that small-holding peasants who increase their agricultural productivity eventually find themselves receiving lower wages from landowners. Increased productivity, if it occurs on plots too small to lead to substantial marketable surpluses, enables large farmers to reduce the level of wages paid to hired workers.[35] Increased agricultural productivity increases the contribution of subsistence plots to the maintenance of workers, allowing the level of wages to decrease, but without leading to a marketed surplus and increased farm income.

We will reintroduce the issue of the capturing of benefits of technological change later in this chapter. Nonetheless, with the transition toward functional dualism based on wage labor provided by smallholders, designers of agricultural development programs are discovering that technological assistance to smallholders, many of whom are more fundamentally wage workers than agriculturalists, may increase aggregate agricultural production in the smallholding sector, but without increasing smallholders' incomes.

133

STATE FISCAL CRISIS AND THE STALLING OF
THE GREEN REVOLUTION

We discussed earlier the political and economic sensitivity of food prices vis-a-vis urban industry. Since high and/or
rising prices engender political resistance to the government
on the part of urban industrialists and workers and threaten to
render export commodities noncompetitive on the world market,
Third World states have tended to pursue policies to minimize
food costs. Two major strategies have predominated. The first
was to subsidize or underwrite the utilization of Green Revolution technologies so as to increase domestic production of
staple foods. These subsidies were usually focused around subsidized inputs or price supports. The second major strategy,
which as we will emphasize later conflicted with the first, was
to maintain overvalued exchange rates (and a lack of protective
tariffs against food imports) so as to depress "artificially"
the costs of imported food commodities.[36]
    Both sets of policies, which were typically pursued in
combination with each other, tended to reduce food costs below
"free-market" levels. However, these policies were also mutually contradictory and costly. For example, the depressing
effect of overvalued exchange rates and food imports on domestic food prices increased the level of subsidies required to
stimulate production of Green Revolution crops. Agricultural
subsidies and overvalued exchange rates also tended to have
adverse effects on the larger economy--diverting scarce government resources from needed infrastructural and social welfare
projects and tilting the economy toward dependence on imports.
    The changing political-economic milieu of the late 1970's
and early 1980s has tended to undermine the simultaneous pur-
suit of Green Revolution subsidies and food imports. First,
the OPEC oil embargo and Iranian oil crisis led to substantial
price increases for energy and energy-related inputs, thereby
increasing the expense of subsidizing the use of petro-chemical
inputs. Second, Third World governments experienced frequent
fiscal crises due to rising energy prices, growing levels of
debt service, and other factors. In response to increased
pressures on government revenues, Third World governments have
tended to find it less expensive to import staple foods than to
produce them domestically through subsidized inputs.[37] As a
result, the "classical" Green Revolution--that is, technological change focused on staple food crops--has begun to lose its
momentum.
    In low-income countries with high levels of land concen-
tration, the early years of the Green Revolution found disproportionate emphasis being placed on inducing technological
change among large farmers, if for no reason other than the
fact that they controlled the bulk of agricultural land. The
Green Revolution thus spread most rapidly among these large
producers. However, by the late 1970's Latin American data

began to show that the estate sector was abandoning the production of Green Revolution staple crops, apparently because of declining subsidies and price ceilings due to the pressures on governments discussed earlier. These large farmers were shifting toward production of agro-export commodities for which the source of demand (developed countries, industry, or affluent middle and upper classes) was more dynamic than the domestic market in staple foods.

Corollary to the trend for large farmers to shift from staple food to agro-export production is the tendency for staple food production to be relegated to favorably situated peasants (relative to smallholders) who produce these commodities with labor-intensive methods (e.g., wheat in Colombia). Staple food production may thus have an uncertain socioeconomic and technological future. These foods, to the degree that they are produced by "middle peasants," will be produced by a declining stratum of the agricultural class structure, as was noted earlier. Second, given the increased emphasis on imports of staple foods, these middle peasants increasingly face the competition of efficient, capital-intensive, large-scale producers in First World countries. Finally, it would not be surprising that the political predominance of large farmers will eventually be translated into alteration of the priorities of national agricultural research institutions, resulting in a shift away from research on staple food crops and toward increased emphasis on agro-export crops.

These changes in the international and national organization of agricultural production, may over the long-term, portend a significant shift in the international division of labor. On one hand, the developed countries (particularly those of North America) may be expected to increase their dominance in the production of cereal grains on a capital-intensive basis, taking advantage of the economies of scale made possible by advanced research on wheat, corn, and rice. On the other hand, Third World countries may become increasingly specialized in the production of labor-intensive (especially agro-export and horticultural) and land-extensive (prototypically livestock) commodities. Most ominous for the purposes of this chapter is the possibility that Third World production may be increasingly tilted away from staple food production and toward production of exports (e.g., groundnuts), luxury commodities (e.g., livestock), and industrial raw materials (e.g., sugar).

These trends, which admittedly are yet nascent but plausible, would present major challenges to research institutions and administrators of agricultural development programs. The most significant challenge would be the deteriorating socioeconomic basis for the production of staple food commodities, with large farmers increasingly oriented toward production for agro-export markets and with peasant producers of staple foods subject to competition with large-scale, capital-intensive farms in developed countries. Through world markets, private eco-

nomic rationale for the adoption of new food production technologies will become less persuasive.

The research and extension techniques now typically termed
"farming system research" will be increasingly necessary to
induce peasants to adopt new technologies. Researchers and
agricultural development administrators will no longer have the
luxury of formulating technological "packages" which represent
a sharp break from traditional production practices.
Technological interventions will have to represent minimal
changes in farming practices and have consequences that
minimize disruption of other components of the "farming
system".[38] However, this mode of research and extension is
quite expensive to put in place across an entire country.
Moreover, one can expect these middle peasant producers of
staple foods to be highly risk adverse[39], and perhaps even more
so in certain respects than near-landless smallholders, leading
to a frustratingly slow pace of technological change.[40]

EMERGENT DILEMMAS IN TECHNOLOGICAL
TRANSFER AND ADOPTION

The previous section of the paper emphasized major socio-
economic changes in Third World agrarian structures and discussed some of the direct implications of these changes for
programs of technology transfer and adoption. In this segment
we want to highlight some further implications of the transformation of Third World agrarian systems and of their changing
roles in the international division of labor.

Technological Change for What Purpose?

It was noted earlier that the locus of capital accumulation in Third World agrarian economies is shifting more square-
ly toward the large farm sector--a sector which appears to be
withdrawing from the production of staple food crops. Further,
the smallholder sector is increasingly becoming a refuge for a
growing rural population against highly inadequate employment
opportunities and in many areas is declining in its status as
producers of marketed agricultural commodities. These transformations are thus in conflict with the basic presumption
behind programs of agricultural R&D: to foster increased productivity which will lead to marketable surpluses of staple
foods, rural capital accumulation, and increased rural incomes.

The emergent dilemma is that ongoing structural trends
dictate growing barriers to capital accumulation through increased production of staple grains. Administrators of agricultural research and extension organizations will thus face an
increasingly transparent and unenviable choice: either to aug-
ment rural capital accumulation by emphasizing technological
improvement in already capital-intensive agro-export commodities generally produced by large farmers, or to emphasize tech-

136

nological improvement in staple food production which will lead to smaller (if not negligible) levels of capital accumulation.

In countries where the middle peasantry has largely disappeared, the dilemma will be even sharper, with technology transfer either centering around enhancing capital accumulation in the large farm sector or focusing on short- to medium-term maintenance of marginalized semiproletarian peasants on fragmented plots. These dilemmas, which reflect a clear choice between addressing the needs of highly polarized social classes, will thus make the processes of technological transfer and R&D increasingly politicized. Politicization, of course, is not inherently undesirable, especially if it enables disadvantaged groups to have greater input into public policy making. However, as we discuss below, the focusing of technological transfer on the smallholder sector does not in and of itself guarantee that smallholders will be the ultimate beneficiaries of technological change.

Capturing the Benefits of Technological Change

It was noted earlier that increased productivity among smallholders may over time result in an indirect subsidy to the large farm sector if these two sectors are connected through wage labor relationships. As a result, many well-meaning alterations of the traditional agendas of the international centers and national agricultural research institutes toward greater attention to smallholders and to farming systems research and extension models may be ineffectual in improving the condition of smallholding peasants.

The dilemma revolves around the crucial fact that if peasant livelihood is to be improved through technological transfer, the peasant must have sufficient land so that surplus food can be produced and marketed, leading to capital accumulation by the peasant. Peasants on subfamily-sized plots, on the other hand, are less able to market agricultural commodities, and given their dependence on wage work, will be unable to improve their level of living if increased productivity reinforces the role of the subsistence sector in partially underwriting the maintenance costs of hired agricultural workers. Technology transfer institutions thus must be aware of the capture issue and recognize the trade-offs involved in working with target audiences on various sized farms.

"Farming Systems" versus Experiment Station Approaches

The agricultural sciences have traditionally embraced the experimental method as their principal research approach, for obvious reasons. The experimental method enables the researcher to isolate the independent effects of "treatments" and to establish the existence and strength of causal relationships. However, there has been considerable criticism of the experiment station approach as applied to technology transfer to Third World countries, especially where scientists have been

137

trained in Western institutions and may be insensitive to the factors that affect peasant decision-making.[41]

One major response to this criticism has been the farming systems approach which seeks to understand the farm as an integrated system of crops, animals, and humans. This approach diversifies the methodologies of agricultural scientists, altering them to the fact that knowledge on how new technologies articulate with the farming system as a whole must be added to experimentally-derived knowledge generated at the experiment station.

Although farming systems approaches have not been without criticism, they do facilitate the diversification of agricultural knowledge and the ability of peasants to utilize new technologies with minimal disruption to traditional farming practices.[42] Farming systems research, however, is labor-intensive--requiring large numbers of multidisciplinary research teams--and hence is expensive, especially for countries with shortages of trained scientists. Moreover, farming systems research results tend not to be a substitute for experiment station research.

The farming systems phase of technology transfer must build on basic research and adapt these findings to specific situations of peasant production. Administrators of technology transfer programs will thus face growing pressures on their budgets, particularly if they elect to place emphasis on the technical needs of smallholders. This dilemma is particularly acute if, as alluded to earlier, the productivity benefits resulting from farming systems approaches cannot be fully or largely captured by the target group.

TECHNOLOGY TRANSFER THROUGH THE INTERNATIONAL
AGRICULTURAL RESEARCH CENTERS: IMPLICATIONS
OF THE GENETIC ENGINEERING REVOLUTION

The international agricultural research centers, funded through the Consultive Group on International Agricultural Research (CGIAR), have spearheaded the major technological changes in Third World agriculture that have occurred over the past three decades. There has been a highly visible and controversial literature debating the costs and benefits of the international centers in the technological and socioeconomic transformations of world agriculture.[43]

In this paper we do not wish to reintroduce these debates, partly because we hope that the foregoing comments have indicated that the institutional context of Third World countries has shifted significantly since the initial surge of adoption of Green Revolution technology. Instead, our concern in this portion of the paper will be with how the role of the international centers might change over the medium- to long-term.

The specific focus of our concern with possible changes in the roles of the international centers related to the emergence of genetic engineering technologies in the advanced industrial societies (especially the U.S.). While there remains a continuing debate as to the extent to which genetic engineering will transform agricultural research and development, there appears to be sound evidence to assert that the next 20 years will witness significant advances in these technologies.[44] Possible technical advances resulting from such research might include biological nitrogen fixation, increased photosynthetic efficiency, pest and disease resistance, and salt, heat, and drought tolerance.[45]

Genetic engineering has massive potential implications for the international centers, since the activities of these centers have historically revolved around traditional plant breeding and the adaptation of agronomic practices to the "needs" of new varieties. Genetic engineering promises to transform--some would say supplant--traditional plant breeding by, for example, greatly improving the speed and efficiency of evaluation of germplasm.

Two characteristics of the emergence of genetic engineering R&D in the advanced industrial societies may render the international centers increasingly peripheral in the process of technological transfer. First, genetic engineering research is extremely expensive due to the costly facilities required for the recombination of DNA, instrumentation, and other genetic engineering techniques. Current international center budgets are clearly too small to allow significant entry into genetic engineering research. Second, the key thrust of U.S. genetic engineering research for agriculture is in the private sector, due, in part, to the expense of this research in a context of stagnant budgets for publicly-funded, nonmilitary R&D. Thus, the research breakthroughs will be largely proprietary and to a degree unavailable to the international centers through the interaction of the U.S. land-grant university, international center, and Third World university research network.

While the point can be overemphasized, the international centers face the prospect of technological obsolescence unless funds provided through CGIAR (or perhaps through U.S. venture capital genetic engineering firms such as the International Plant Research Institute or Cetus) are greatly increased. To the degree that the publicly-funded agricultural research network experiences technological obsolescence because of private sector dominance in genetic engineering, the trajectories of technological change in the Third World will be increasingly dictated by proprietary interests in the marketing of genetic engineering-related inputs. More specifically, the international centers and their Third World research institute affiliates may lose control over the development of state-of-the-art technologies to be deployed in Third World contexts, and research on promising but non-commercializeable technologies may be eschewed by genetic engineering research firms in the U.S. and other developed countries.

These comments are not to suggest that genetic engineering technologies will be adverse for Third World farmers. Indeed, genetic engineering may lead to technologies such as bacterial pesticides, pest and disease resistance, or plant responsiveness to low levels of fertility that are better suited to Third World conditions than the classical Green Revolution packages. The point we wish to make is that the genetic engineering frontier may decisively change the terms of the technological change debates that have been raging for over a decade.

The rise of private sector dominance in advanced agricultural R&D may lead to recognition of the potential importance of public research institutions and of the desirability of a major public role in technology R&D. Contemporary critics of the international centers may thus ironically find themselves in coalition with these and other publicly-funded institutions to create structures for the representation of the "public interest" in the formulation of research priorities.

DISCUSSION

We began this paper with a brief reflection on the evolution of the international agricultural development literatures over the past few decades. In a sense we have premised this paper on detailing the implications for agricultural development practice of several literatures--especially political economy and several variants of "peasant studies"[46]--which have infrequently been brought to bear on applied development issues. Reflecting again on the evolution of development theory and practice, a paper focusing on the topic of barriers to technological transfer and adoption written only 20 years ago would no doubt have placed principal emphasis on peasant traditionalism an other cultural constraints on modernization and technological change.

Our implicit emphasis throughout this chapter has been on structural and organizational rather than critical barriers to socioeconomic and technological progress. As Barlett has demonstrated, peasants tend to be rational decision-makers and to make choices that maximize their welfare, given the socioeconomic constraints they face.[47] We do not wish to deny the importance of cultural forces in agrarian societies (e.g., religious beliefs, ethnic antagonisms). Moreover, it is crucial for agricultural development practitioners to understand prevailing patterns of sexual division of labor in both their material and cultural aspects.[48] Nevertheless, it is important for development practitioners to understand the global and national dynamics of the changing international division of labor, particularly if they are concerned with the long-range planning of technology transfer.

To some this chapter will seem excessively pessimistic and as overemphasizing constraints and barriers rather than opportunities and potentials.[49] In a certain sense it is cautiously pessimistic; we see a number of barriers that are deep structures in the world economy. International political structures will mitigate against rapid, equitable transfer of technology to (or development of technology within) low-income countries.

This pessimism must be tempered, however, by the rapid strides that have been made in U.S., multinational, and Third World national development institutions to be more realistic about the role that technology transfer can play and to take more seriously the technological needs of peasant agriculturalists. A great deal remains to be done, but the progress that has been made is considerable.

REFERENCES

[1]Yujiro Hayami and Vernon W. Ruttan, Agricultural Development (Baltimore: Johns Hopkins University Press, 1971).

[2]Douglas S. Pauw and John C. H. Fei, The Transition in Open Dualistic Economies (New Haven, Conn.: Yale University Press, 1973).

[3]Andre Gunder Frank, Dependent Accumulation and Underdevelopment (New York: Monthly Review Press, 1978).

[4]John Taylor, From Modernization to Modes of Production (Atlantic Highlands, N.J.: Humanities Press, 1979).

[5]James C. Scott, The Moral Economy of the Peasant (New Haven, Conn.: Yale University Press, 1976).

[6]Goran Hyden, Beyond Ujamaa in Tanzania: Underdevelopment and an Uncaptured Peasantry (Berkeley, Calif.: University of California Press, 1980).

[7]Samuel L. Popkin, The Rational Peasant (Berkeley, Calif.: University of California Press, 1979).

[8]Frederick H. Buttel, "Energy Agrarian Structure, and Food Production," Cornell Rural Sociology Bulletin 122 (August 1981). Some may take exception to our view of the international division of labor as the basic force. This may point to a group of policies which adversely affects rural areas and makes technological transfer and adoption of agricultural practices in the rural areas difficult. We, however, tend to view these policies as part of the results of the international division of labor. See World Bank, Accelerated Development in the Sub-Saharan Africa: An Agenda for Action (Washington, D.C.: The World Bank, 1981), Chapter 2, Basic Constraints.

[9]Albert O. Hirschman, The Strategy of Economic Development (New Haven, Conn.: Yale University Press, 1958); W.W.

Rostow, The Stages of Economic Growth (Cambridge, Mass.: Cambridge University Press, 1963).

[10]Alain de Janvry, "Nature of Rural Development Programs: Implications for Technology Design" in Economics and the Design of Small Farmer Technology, eds. A. Valdes et al. (Ames, Iowa: Iowa State University Press, 1978).

[11]Folker Frobel, The Current Development of the World Economy (Tokyo, Japan: United Nations University, 1980); Peter Evans, Dependent Development (Princeton, N.J.: Princeton University Press, 1979).

[12]Theodore W. Schultz, ed., Distortions of Agricultural Incentives (Bloomington: Indiana University Press, 1978).

[13]Michael Lipton, Why Poor People Stay Poor: Urban Bias in Rural Development (Cambridge, Mass.: Harvard University Press, 1977); Robert H. Bates, Markets and States in Tropical Africa: The Political Basis of Agricultural Policies (Berkeley, Calif.: University of California Press, 1981).

[14]Alain de Janvry, The Agrarian Question and Reformism in Latin America (Baltimore: Johns Hopkins University Press, 1981).

[15]Rosemary E. Galli, ed., The Political Economy of Rural Development: Peasants, International Capital, and the State (Albany: State University of New York Press, 1981).

[16]Ian Roxborough, Theories of Underdevelopment (Atlantic Highlands, N.J.: Humanities Press, 1957).

[17]Paul A. Baran, The Political Economy of Growth (New York: Monthly Review Press, 1957).

[18]Andre Gunder Frank, Latin American: Underdevelopment of Revolution? (New York: Monthly Review Press, 1969).

[19]Evans, Dependent Development; Roxborough, "Theories of Underdevelopment;" Alain de Janvry and Louis Crouch, "Beyond Dependency: New Directions in Latin American Political Economy," Department of Agricultural Economics, Working Paper no. 96 (Berkeley, Calif.: University of California, 1980).

[20]Immanuel Wallerstein, The Capitalist World-Economy (Cambridge, Mass.: Cambridge University Press, 1979).

[21]Evans, Dependent Development.

142

[22]Roxborough, Theories of Underdevelopment.

[23]Peter Dorner, Land Reform and Economic Development (Baltimore: Penguin, 1972).

[24]R. Albert Berry and William R. Cline, Agrarian Structure and Productivity in Developing Countries (Baltimore: Johns Hopkins University Press, 1979).

[25]Ibid.

[26]de Janvry, The Agrarian Question; Donald K. Freebairn, "Review of Berry and Cline, Agrarian Structure and Productivity in Developing Countries," Rural Sociology 46 (Winter 1981): 745-77.

[27]de Janvry, The Agrarian Question.

[28]Keith Griffin, The Political Economy of Agrarian Change (Cambridge, Mass.: Harvard University Press, 1974); Andrew Pearse, Seeds of Plenty, Seeds of Want (New York: Oxford University Press, 1980); Kenneth A. Dahlberg, Beyond the Green Revolution (New York: Plenum Press, 1979).

[29]de Janvry, The Agrarian Question; Galli, "The Political Economy."

[30]"Full proletarianization," by contrast, would imply full loss of access to land combined with exclusive reliance on wage income for family support. De Janvry argues that full proletarianization would lead to higher wage equilibria than semi-proletarianization, since the maintenance of a fully proletarianized labor force would require wage levels at or close to the level necessary to reproduce the household unit.

[31]Bryan Roberts, Cities of Peasants (London, England: Edward Arnold, 1978).

[32]Peter Peek, "Agrarian Change and Rural Emigration in Latin America," ILO Working Paper (Geneva, Switzerland: ILO World Employment Programme Research, International Labor Organization, 1978).

[33]Dale L. Johnson, "Observations on Rural Class Relations," Latin American Perspectives 9 (1982): 2-11; James D. Cockroft, Mexico: Class Formation, Capital Accumulation, and the State (New York: Monthly Review Press, 1982).

[34]Peek, "Agrarian Change."

143

[35]M. Taussig, "Peasant Economies and the Development of Capitalist Agriculture in the Cauca Valley, Colombia," Latin American Perspectives 5 (1978), pp. 62-91; Galli, "The Political Economy."

[36]Bates, "Markets and States."

[37]Luis Crouch and Alain de Janvry, "The Class Basis of Agricultural Growth," Department of Agricultural Economics, Working Paper no. 70 (Berkeley, Calif.: University of California, 1979).

[38]Patricia Garrett, "Farming Systems Research: An Appreciation and a Critique" (Paper presented at the annual meeting of the Rural Sociological Society, San Francisco, September 1982); Richard R. Harwood, Small Farm Development (Boulder, Colo.: Westview Press, 1979); R.E. McDowell and P.E. Hildebrand, Integrated Crop and Animal Production: Making the Most of the Resources Available to Small Farms in Developing Countries (New York: The Rockefeller Foundation, 1980).

[39]Peggy F. Barlett, "Adaptive Strategies in Peasant Agricultural Production," Annual Review of Anthropology 9 (1980): 545-973.

[40]Frank Cancian, "The Innovator's Situation (Stanford, Calif.: Stanford University Press, 1979).

[41]Barlett, "Adaptive Strategies;" William S. Saint and E. Walter Coward, Jr. "Agriculture and Behavioral Science: Emerging Orientations," Science 197 (1977): 733-737; Dahlberg, "Beyond the Green Revolution."

[42]Garrett, "Farming Systems Research."

[43]Griffin, "The Political Economy;" Michael Perelman, Farming for Profit in a Hungry World (Montclair, N.J.: Allanheld, Osmun & Co., 1977); Sterling Wortman and Ralph W. Cummings, Jr., To Feed This World (Baltimore: Johns Hopkins University Press, 1978); David Pimentel and Marcia Pimentel, Food, Energy, and Society (New York: Wiley, 1979); Dahlberg, "Beyond the Green Revolution;" Richard W. Franke and Barbara H. Chasin, Seeds of Famine (Montclair, N.J.: Allanheld, Osmun & Co., 1980).

[44]Martin Kenney, Frederick H. Buttel, J. Tadlock Cowan, and Jack Kloppenburg, Jr., "Genetic Engineering and Agriculture: Exploring the Impacts of Biotechnology on Industrial Structure, Industry-University Relationships, and the Social Organization of U.S. Agriculture," Cornell University Bulletin 125 (July 1982); Kenneth O. Rachie and Judith Lyman, eds.,

Genetic Engineering for Crop Improvement (New York: The Rockefeller Foundation, 1981).

[45]It should be noted, however, that genetic engineering techniques in conjunction with industrial microbiology may result in the undermining of agro-export production in Third World countries. For example, the application of these to produce the sweetner aspartame and high fructose corn syrups (through immobilized enzyme processes) could effectively displace sugar, one of the major export crops produced in low-income countries.

[46]Much knowledge of potential relevance to agricultural development practice is presently being reported in the Journal of Peasant Studies, but this journal has not been widely circulated within U.S. agricultural development circles because it is published in England and prints relatively few articles by U.S. scholars.

[47]Barlett, "Adaptive Strategies."; Caroline Hutton and Robin Cohen, "African Peasants and Resistance to Change: A Reconsideration of Sociological Approaches," in Beyond the Sociology of Development, eds. I. Oxaal et al. (London, England: Routledge & Keegan Paul, 1975). Hutton and Cohen present an excellent theoretical discussion of "peasant traditionalism" and effectively argue that the failure of peasants rapidly to adopt modern technologies usually reflects a realistic posture toward risk and other constraints.

[48]Carmen Diana Deere and Magdalena Leon de Leal, "Peasant Production, Proletarianization, and the Sexual Division of Labor in the Andes," Signs: Journal of Women and Culture in Society 7 (1981):338-360.

[49]For a positive review of the accomplishments of technological transfer and development assistance see John P. Lewis, "Development Assistance in 1980's" in U.S. Foreign Policy and the Third World Agenda 1982. eds. Roger D. Hansen et al. (New York: Praeger, 1982).

10
Host Country Institutions and Diffusion of Technology

George H. Axinn

Host country institutions should be taken into account in planning strategies for transfer of food production technology to developing nations. Too often strategies have been developed for technology transfer based on various technological considerations, ignoring the economic, the social and the cultural considerations--to say nothing of host country institutions.

The recognition that host country situations exist is an important first step. They play a vital role in the introduction of new technology or the transfer of old technology. Normally, such institutions act to protect host country people from inappropriate technology which some "outsiders" may wish to introduce. Thus their role is crucial.

This chapter touches on four considerations with regard to these institutions. First, I will deal briefly with indigenous knowledge systems and indigenous learning systems. Then, I will point to the key role of host country institutions in finding technology which "fits." Third, I will deal with the issues of preparing a rural social system for technical change. And finally, I will make some suggestions about reciprocity in intersystem interactions.

INDIGENOUS LEARNING SYSTEMS

Wherever in the world you might go, indigenous knowledge systems already exist in that place.[1] Also, indigenous learning systems will be discovered by the careful observer. Their function is to develop new knowledge and to transfer and disseminate knowledge among the people of that system. Whether the concern is the production of food, health care, transportation, or any other aspect of life, there will be technologies within the rural social system for handling those matters. These knowledge systems exist. They are important. Failure to recognize them will be costly, if not defeating, when it comes to introduction of new technologies. Thus, the strategy advocated here is to discover them first, and then build alliances with them. The alternative is to be in competition, if not in conflict.

With regard to the host country institutions, from my perspective there are at least two general types. One is the "insiders" institution. The other is the "outsiders" institution.

The "insiders" institution is based on the values, the knowledge systems, the structure, and the power system of the local group. Some examples are institutionalized arrangements for the construction of such farm buildings as grain storages or livestock barns; local systems for threshing of harvested crops; or local systems for cooperative work in planting.

It is normal, for example, for there to be institutionalized patterns for planting any particular crop. These will include the patterns of who has the responsibility of allocating last year's crop, perhaps, between consumption and storage for seed. For the planting exercise itself, who makes the depressions in the soil and who inserts the seed, etc.[2] Quite often these are gender-sensitive phenomena, in which there are prescribed roles for men and for women. Beyond these, there are usually group institutions concerned with such things as the accumulation of savings, or the preservation of territory.[3]

The "outsiders" institutions have names like our own. They may be agricultural research institutions, agricultural schools, or agricultural extension systems. They tend to have been introduced by people from outside social systems, and they tend to reflect the values, norms, and operational procedures of other places in the outside world.

Introduction of new technology into such a system is like transplanting an organ into a human being or some other animal. Two steps are critical. One is finding an organ which actually **fits** the system in which it is to be introduced. The second is preparation of the recipient organism so that it will be ready to receive the transplant. Both are paralleled in the exercise of technology transfer.

FINDING TECHNOLOGY WHICH FITS

For this discussion, technology is defined as a body of knowledge applied to specific ends. The "fit" is related to the specific ends.

The problems of "fit" are overwhelming. Normally, technologies developed by application of the science of one society to the problems of that society are likely to be technologies which are uniquely appropriate for use in that society. Using somebody else's technology is like using somebody else's toothbrush. In our culture, we simply don't do it. The science behind it may be shared. The idea of keeping one's teeth clean and free of food particles may be sound anywhere. But we simply don't feel it is appropriate to use each other's toothbrush. Using somebody else's technology is a similar phenomenon.

In agriculture, in particular, some technology is developed for large-scale farming systems; other people have small-
scale farming systems. Some technology is developed for highly
specialized, monocrop farming systems; while other technology
is appropriate for unspecialized, mixed, multi-crop and livestock farming systems. Some technology is appropriate for high
energy farming systems; other technology is appropriate for low
energy farming systems. Some technologies are appropriate for
capital-intensive farming systems; other technologies are
appropriate for low capital agricultural systems. Some
technology is appropriate where there is a surplus of land; a
different technology tends to be appropriate where there is a
shortage of land. Some technology is appropriate when there is
a shortage of human labor; a different technology is
appropriate when there is a surplus of human labor.[4]

Therefore, it would be a very rare event when technology
developed for large-scale, specialized, energy-intensive, capital-intensive, agriculture with a surplus of land and a shortage of human labor, would be appropriate for the other kind.
What I have just described is typical of American contemporary
agriculture. The former is found on the small mixed farming
systems of much of Asia, Africa, and Latin America. Since
there is a tendency to find small-scale, mixed, low-energy,
low-capital farming systems with a shortage of land and a
surplus of labor, it is indeed rare when technology from one of
these types of systems is suitable for transfer to the other.

There are many examples of attempts to transfer technology
which does not fit. I am not pointing a finger at anyone, being one of the "perpetrators" of such transfer myself. I was
involved in bringing tractors to the small farming systems of
what is now called Bangladesh. How that was possible, reflecting on what I have just said, is hard to understand. I have
also worked with others bringing chemical insecticides to very
small farming systems, where the practical thing to do when you
have an infestation of some kind of insect is to take the whole
family into the field and pick them all off by hand. If you
are in a culture which does not normally consume such insects,
you can just squash them and leave them. In much of the world
the practical thing to do is to eat them.

And we have seen the same kind of phenomena where short-
stem cereal crops are introduced in an area where large numbers
of ruminant livestock are absolutely dependent on the straw
from the cereal crop. Such a technology introduction fits very
nicely in social systems where cereal crops are grown only for
the grain. It certainly does not fit in areas where the same
crop is grown both for the roughage for ruminant livestock and
for the cereal grain.[5]

In all of this, there is a key role for host country institutions. They can take the science of the so-called
"western world" and mix it with the indigenous science, and out
of that mixture may evolve technologies which "fit." They do

148

not always do this. Sometimes their values, particularly their academic and scientific values, have been so warped by training in the outside world that they, too, have difficulty knowing what fits and what doesn't. Farming systems research is an attempt to solve these problems.[6]

PREPARING A RURAL SOCIAL SYSTEM
FOR TECHNICAL CHANGE

The preparation of the recipient system is a highly complex matter. Even if a particular technology does fit, who in the local rural social system will benefit if it is successfully introduced? Who will suffer if it is successfully introduced? Again, recent attention to the problems of women in international development has illustrated cases where the benefits of technological change are gender sensitive.

The fishing village, which converted from small-scale hand-operated drying arrangements to a gas fired dryer for fish is an example. The village was able to increase its total production significantly. The men of that village who did the fishing were able to bring in larger quantities of fish and realize greater profits from their work. The women of the same village who formerly dried the fish, on the other hand, were literally out of a job—and disadvantaged both economically, sociologically and psychologically.[7]

Similarly, some of the technological introductions which have featured economies of scale, have resulted in a change in landlord-tenant relationships in which the landlord literally throws his tenants off the farm, so that the whole unit can be managed as one. This often results in greater productivity, and certainly greater wealth for the land owner. The consequence for those who tilled that soil before the technology change is usually a condition known as landlessness—and tends to result in poverty and migration to urban centers.

From the perspective of host country institutions—particularly what I have called the "outsiders institutions"—those concerned with agricultural research and extension education—there are similar questions. Which members of which staffs will be given the credit for this technology? Who will be blamed for this technology if it fails? If successfully introduced, then who gets the credit? Who takes the blame?[8]

Among host country institutions, there are, of course, others besides those concerned with agricultural extension or research. For every function in a rural social system, there are one or more institutions to meet the need. These are valued in any host country and people will defend them and protect them. For any particular technological innovation, they may be either friends or enemies—depending on the nature of the innovation, and the extent to which its consequences are predicted to be favorable. All of these functions are related to one another. Any change in any one of them affects the whole system. Therefore, diplomacy and negotiation with such other host country institutions may be critical--both for the introduction of new technology and for the sustained use of that new technology after outside introducers have gone.

The functions to which I am referring here--and which need to be taken into account--include not only the function of production--producing the food and fiber--particularly here producing fish products--but also the supply of inputs to such production and the marketing of the outputs from that production. Marketing could be merely consumption by the producers-- or it could be marketing through a food chain to outside systems.

Beyond the three economic functions of production, supply, and marketing, every rural social system has at least four other functions. These include governance, or the control of the other components; personal maintenance within the system, by means of which human beings feed, clothe, and house themselves; health care delivery within the system, which is always part of the indigenous apparatus; and the learning functions which exist in every rural social system--and which include both education and research. All of these, in turn, are part of a larger social, economic, political, cultural, geophysical, and agroclimatic set of considerations. Failure to take them all into account--including the interactions among them and the institutions designed to protect each one of them, can mean failure in the introduction of technology which otherwise is quite appropriate.[9]

RECIPROCITY IN INTERSYSTEM RELATIONS

Finally, let us consider reciprocity between host country institutions and outside institutions. There are difficulties with any system in interaction. These include the interaction between "outside" institutions and "inside" institutions-- between, for example the USDA and the host country ministry of agriculture; or the U.S. university and the host country university; or the international agricultural research institute and the host country agricultural research system.[10]

There are a whole series of constraints which can be taken into account when planning transfer of food production technology. Every single piece of technology can only be transferred from one system to another by human beings. Hence, there is always a human constraint. Human beings from one system, interacting with human beings from the other systems, tend to have spouses, children, parents, health problems, food consumption patterns, and other related human constraints, which if left unattended, tend not to enhance the technology transfer but to interfere with it.

Similarly, there are always cultural constraints. People happen to have different languages, different religions, different values. Failure to take the cultural constraints into account can again serve to defeat attempts at technology transfer.

Beyond these, both in the host country institution and in the outsiders' institutions, there are always administrative arrangements. These relate to the logistics of the transportation, communication, financial support, and a variety of other things. Typically the administrative arrangements of the host country institution will be different from the administrative arrangements of the outsiders' institutions. Again these differences can be a constraint.

Beyond the administrative, inevitably, there are political constraints. Both systems will have their own political considerations--the rules and regulations of the political game, plus the dynamics of a changing power structure. These, again, are always a constraint on the system.

And if you want to include both the host country institution and the outsiders' institutions, there are always intersystem diplomatic constraints which affect the system.

CONCLUSION

Host country institutions are different from U.S. institutions, even when they carry a similar name. In the USA, the agricultural extension system is heavily controlled by the people it serves, particularly at the county level. In most countries there is an agricultural extension system, but it is controlled in the central capital city, not by its clients. Similarly in agricultural research, North Americans tend to think of agricultural experiment stations which are responsible to organized farmer commodity groups. In most other countries, farmers are not organized, have little clout when it comes to agricultural research, and are usually not consulted. Besides these differences, so many host country institutions and their staffs are young and inexperienced and are susceptible to domination by foreigners, or domination by those with influence.

Host country institutions can be treated like professional partners, with equal confidence, concern, and responsibility; or they can be mistreated. When the outsiders consider themselves superior, the insiders know it. The superiority syndrome, a "disease" too often found among professionals trying to transfer their technology, is a serious problem. The superiority syndrome alone can defeat the transfer of food production technology, even where it fits technically.[11]

But when full partnerships develop--between host country institutions and outside institutions which share a mutual interest; when neither is forced to be either donor or recipient; when each looks forward to the benefits to themselves; when both are aware of the costs to themselves; when host country institutions and outsiders' institutions develop an iterative reciprocity--so that through a series of exchanges each can weigh the costs and benefits, and find that it is in their

151

interest to continue the relationship--then, together, the achievements can be great.[12]

Together they can develop technologies which fit. Together they can enhance the adoption of appropriate technologies. Together they can contribute to the improvement of the human condition.

REFERENCES

[1]David W. Brokensha, D. M. Warren, and Oswald Werner, Indigenous Knowledge Systems and Development (Lanham, Maryland: University Press of America, 1980).

[2]Marvin Harris, Cannibals and Kings: The Origins of Cultures (New York: Vintage Books, 1978, Random House, Inc., 1977).

[3]See for example: Irene Tinker, Women and Energy: Program Implications, mimeographed (Washington, D.C.: WID/USAID, 1980); Ester Boserup, Women's Role in Economic Development (New York: St. Martins Press, Inc., 1970); Elise Boulding, The Underside of History - A View of Women Through Time (Boulder, Colorado: Westview Press, 1976); Sylvia A. Chip and J. J. Green, eds., Asian Women in Transition (University Park, Pennsylvania: Pennsylvania State University Press, 1980); Roslyn Dauber and M. Cain eds., Women and Technological Change in Developing Countries (Boulder, Colorado: Westview Press, 1980).

[4]George H. Axinn and Nancy W. Axinn, Social Impact, Economic Change, and Development--With Illustrations from Nepal, Michigan State University Farming Systems Research Group, Working Paper no. 13 (East Lansing: Michigan State University, 1981).

[5]George H. Axinn, New Strategies for Rural Development (East Lansing, Michigan and Kathmandu, Nepal: Rural Life Associates, 1978), Chapter 16.

[6]Beverly Fleisher and George H. Axinn, "An MSU Approach to Farming Systems Research," Michigan State University Farming Systems Research Group, Working Paper no. 10 (East Lansing: Michigan State University, 1981).

[7]Irene Tinker, "Energy Needs of Poor Households," Working Papers on Women in International Development no. 4 (East Lansing: Michigan State University, 1982).

[8]William F. Whyte, Participatory Approaches to Agricultural Research and Development: A State of the Art Paper, (Ithaca, N.Y.: Cornell University, Rural Development Committee,

1981;) Normal T. Uphoff and Milton J. Esman, Local Organization for Rural Development: Analysis of the Asian Experience (Ithaca, N.Y.: Cornell University Rural Development Committee, 1974).

[9]Axinn, New Strategies for Rural Development, p. 17.

[10]George H. Axinn, Toward a Strategy of International Interaction in Non-Formal Education (East Lansing, Michigan: Michigan State University, Program of Studies in Non-Formal Education, 1974).

[11]Axinn, New Strategies for Rural Development, p. 36.

[12]Ibid.

11
Irrigation Development: Technology Traditions and Transfers

E. Walter Coward

INTRODUCTION

Presently a major technological solution for increasing food production and alleviating poverty, particularly in the Asian region, is irrigation development. Irrigation development includes both the creation of new irrigation systems and irrigated areas, as well as the improvement of existing irrigation systems. Large portions of the agricultural and rural development budgets of national governments and international agencies are being allocated to irrigation development. And even larger investments are being recommended.

In contrast to the euphoric calls for greater investments in irrigation development are the realities of past irrigation projects that have required unexpectedly long periods for completion, projects that have created far less irrigated area than planned, and cropping patterns and crop yields far different from those anticipated by central policy makers. In some instances, irrigation development has created serious environmental problems and resulted in the actual reduction of productive lands.

Typically, the solution to these problems of irrigation development have been sought in the realm of better irrigation technology: lined canals to replace earthen channels, more efficient pumps to lift water, better control structures to regulate water flows. More recently, it has been suggested that many of these problems are the consequences of inadequate organizations and institutions, and that solutions are to be found through improving the social organization of irrigation. At least two components are emphasized: the organization of the water users and, in those irrigation systems where it operates, the organization of the irrigation agency.

The more satisfactory view, I believe, is the one that recognizes the strong interactions between technology and the social organization--both the organizational imperatives that may attach to various technologies and the limits to technology utilization that may be associated with particular sociocultural forms. As with many other elements of agricultural development, irrigation development is likely to proceed most

effectively when attention is given to these technology-organi-
zation interactions.

This chapter will be focused on issues of irrigation
development in the Asian region both because of its high importance there as well as the fact that this is the region with
which I am most familiar.

IRRIGATION DEVELOPMENT IN ASIA

Perhaps the most important trend in irrigated agriculture
in Asia is towards more intensive use of the irrigation systems: higher cropping intensity, higher use of modern inputs,
and a consequent demand for water to be provided reliably, pre-
dictably, and flexibly. As these forces compel more concern
with the effective performance of irrigation systems, there is
a related need to reconsider the matter of the organization of
the irrigation sector and the present structure of irrigation
investment. In particular, the suitability of continuing to
concentrate the governance of irrigation investments in state
bureaucracies must be carefully examined.[1] Central aspects of
good system performance--for example, reliability and flexibil-
ity--have been elusive bureaucratic achievements. Furthermore,
the inability to induce local investments in irrigation development reduces the rate of irrigation growth and often acceler-
ates the deterioration of past investments.

The phenomenon of irrigation organized under the command
of large-scale bureaucracies, with the specialized role of creating and operating irrigation system, is a relatively recent
phenomenon in Asia, Wittfogel, to the contrary notwithstand-
ing.[1] It is largely associated with the intrusion of the colonial powers into the agricultural activities of the region.
Prior to that period, while various petty kingdoms may have
provided a supportive setting in which permanent wet-rice agriculture could be developed, the actual creation and governance
of irrigated areas was the domain of local groups.[2] And, as
discussed below, these local systems remain important throughout Asia at the present time, as well.

The colonial powers entered an irrigation scene very different from that which they left behind. The pre-colonial
arrangement in much of Asia was composed of numerous small-
scale local level systems which, while they may have been under
the general purview of some regional political system, were not
governed by a state technical bureaucracy. These units represented local investments, sometimes with the aid of regional
elites and subject to their extractions, but essentially local
investments, locally controlled.[3]

An interesting question that arises is why the colonial
powers did not work through these community systems, to pursue
their objectives (a question that might also be posed to current foreign donor agencies). While it is true that much of
the colonial era irrigation development focused on new areas

155

(such as the North coast of Java or the Mekong Delta in Viet-nam), the policies of development neither encouraged community investment or governance in these new areas nor gave significant support to the existing community systems.[4] It is significant to note that the colonial administrations had increases in the production of export (or at least non-food) cash crops rather than subsistence food crops, as a major objective.[5] A major problem they faced in increasing such production was acquiring the necessary labor pool (for their non-irrigated crops as well).

Lack of interest in supporting the community systems, in part, might have derived from the need to "encourage" the move of cultivators into the newly developed, and state-controlled, irrigation perimeters. As cultivators moved into these new facilities, the colonial administration thereby gained further control over them by managing access to a strategic resource--water. Often, colonial administrations faced the need to exert power through a limited staff; thus, there frequently was the urge to concentrate the scattered local population and exploit them through state control of access to some critical resource--everything from access to the supernatural to access to land and water.

Irrigation development in much of Asia has been highly distorted by a process of state concentration of investments and governance and the commitant demise of local rights and initiatives. The state control of irrigation, of course, was consistent with, and part of, the larger sphere of state control of agriculture. Central irrigation control is organized on a series of principles that give prominence to the state in irrigation investment and governance. These include the state's ownership of water, its right to taxation of surplus, its responsibility to invest in water control, and its right to management authority through a technical bureaucracy.

Somewhat remarkably, this pattern has been perpetuated by the new states and most of the influential aid agencies. To a large extent, this seems to result from viewing irrigation development largely (perhaps exclusively) as a technical engineering problem. From this, it follows that scarce technical expertise is best located in a powerful state bureaucracy with a critical mass of well-trained engineers. The technical staff of the donor agencies, likewise, find this locus of expertise efficient for their program objectives. As will be discussed here, there is reason to believe that this approach is increasingly ill-fitted to contemporary needs.

Some might argue that the Japanese experience with irrigation development seems inconsistent with the above ideas since in Japan, even without the colonial experience, there has been a trend toward greater concentration of state investment and governance. But, there are several points to note that counter this argument. First, while many contemporary Japanese irrigation systems are large-scale in command they have been created from an aggregation of smaller, preexisting systems.[6] Second, while there is a large role for the bureaucracy, no doubt related to the state's interests in the heavy subsidies which it provides, there continues to be a very dynamic tension between this bureaucracy and the strong local-level irrigation groups regarding the control and management of irrigation.[7] Third, a significant portion of the Japanese investment in irrigation infrastructure development has been through subsidies and low-interest loans to local groups, such as the land improvement associations; thus, inducing local investment and preserving local control.[8]

The Dual Structure of Irrigation

In many Asian countries the fundamental structure of the irrigation sector is a "dual" one; a government sector in which irrigation development is highly concentrated in the control of a state bureaucracy and a community sector composed of a large number of small, independent, and locally-controlled entities. Referring to this structure as a dual one is not meant to imply that one sector (the community one) is stagnant, "traditional," or inefficient while the other has the opposite characteristics. Rather the distinction is made to draw attention to the two components of the sector, and to suggest that they have a number of very different features, and that their needs for assistance may be divergent.

Recognition of this duality and its origins suggests several large policy questions that will require attention during the 80's and beyond:

> What should be the policy toward the perpetuation of this dual structure?
>
> What, if anything, should be done to assist community irrigation systems?
>
> What can be done in large agency-managed systems to restructure investment and governance to increase farmer involvement?

Perpetuating the Dual Structure

Much thinking about eliminating the dual structure begins with the perspective that the community sector will be engulfed by the government sector. That is, the presently autonomous community systems eventually will be "upgraded" and "improved," and subsequently managed and controlled by the state bureaucracy.[9] Such a strategy is highly suspect from at least two perspectives (1) it assumes that the state bureaucracies have the capacity to take on this additional responsibility, and (2) it implies that the community systems presently are being poorly managed and maintained.

157

Taking the latter assumption first, there is now considerable evidence that many community systems are well-managed and that many more have the capacity to be so with some assistance from regional authorities.[10] Regarding the first point, the uneven capacity of many irrigation bureaucracies should raise considerable concern about policies designed to significantly expand their range of responsibilities--particularly their assuming management of a number of geographically dispersed and individually distinct community systems.[11]

As Tamaki has suggested, the centralization of state control of irrigation should neither be seen as inevitable nor desirable.[12] Policies favoring the perpetuation of this dual irrigation sector can be based on the positive assessment that existing local irrigation organization is a strategic resource on which to build and through which to invest and not simply to be replaced by the state irrigation agency. As with other strategic resources, one seeks to invest in local irrigation organization rather than ignoring or attempting to replace it (with what often is a more expensive and less effective resource).

Lastly, any positive policy to perpetuate the dual irrigation sector should avoid neglecting the community irrigation sector. While we do not yet have a coherent set of policies and strategies for assisting the community sector, there are important cases from which a considerable amount can be learned. For example, the Indonesian program of village development subsidies provides many communities with locally-controlled funds which they are able to invest in rehabilitating or elaborating their community irrigation works.[13] This program rests on community understanding of the irrigation problems and local priorities as to what requires primary attention. These external subsidies have usually been combined with locally mobilized resources, both money and manpower, thus expanding the total community investment in irrigation infrastructure.

In some cases, these investments have been of limited benefit because the community has lacked certain technical information, and inappropriate or poorly designed structures have been built. There is the potential for government to assist with these matters, but only if government itself is prepared to recommend and design a technical apparatus that fits the situation.

And an important part of that fitting is the choice of technology that supports community autonomy and self-reliance, and avoids or minimizes dependency on an outside state agency. As discussed below, this may be the most fundamental principle to be applied in designing assistance for community systems. Technical assistance should not be provided at the cost of increased dependency.

ASSISTING COMMUNITY SYSTEMS

In Southeast Asia, there is a very clear trend for governments to provide more assistance to existing community irrigation systems--the subaks of Bali, the zanjeras of the Philippines, or the muang-fai systems of northern Thailand.[14] There is a very real danger that this "assistance" is being formulated in the narrow terms of technological improvements which are neither consistent with the existential realities of these systems nor with the deeper aims of irrigation development.

Assistance to these community systems often is highly arrogant, seldom respectful. This arrogance usually is the product of the outsider's assumption of superior knowledge and technical skills, combined, most often, with an absence of information about, and understanding of, what is happening in these systems. Thus, as with the famous subaks of Bali, there is little understanding of, or appreciation for, the intricate manner in which water is apportioned to the user, or the refined manner in which users are motivated to sustain the physical apparatus of the system over long periods of time. Thus, it is unsurprising though not less distressing, that external planners should view their assistance to subaks as simply one more instance of constructing a better dam or modern water control structures.

Assistance to community systems also is often distorted because the long-range aim of creating a stable irrigated sector is displaced with the short-term project goals of constructing a dam or building "better" tertiary-level facilities.

There are, thus, serious negative consequences that can result from the encounter between a community irrigation system and an external irrigation development agency.[15] A key issue of the 1980s will be devising policies and procedures to assist without ravaging. The outlines of such approaches are not that difficult to discern: what is uncertain is whether or not the prerequisites of such a a progressive policy can be met; these include: the willingness of highly centralized governments to support autonomous local entities; the ability of the technical bureaucracy to retrain its professionals and/or hire more staff with appropriate backgrounds; agency interest and capacity to formulate needed new procedures.

One approach may be to establish specific support systems to work with community systems-support systems that would have different goals than the arm of the agency concerned with operating large-scale systems, perhaps a different mix of professional staff, and a new mode of relating to the community. A prototype of this support system is emerging in the province of North Sumatra in Indonesia where the provincial and district offices of agriculture are adding to their staff junior engineers who will be able to provide simple technical advice to communities engaged in improving their irrigation works. Among the tasks that such support systems might perform are: techni-

159

cal advice on system operation, planning and design assistance, provision of credit and subsidies for improvements and system elaboration, help with securing water rights, and aid in dispute settlements.[16]

Implementing these strategies ultimately must rest on the principle that government can, and will, complement local initiatives without controlling. To do otherwise with regard to community systems is to risk the great danger that the attempt to improve this sector will, in fact, lead to its collapse. Assistance to community systems, unless carefully staged, can encourage the demise of community investment and responsibility, and attach the fate of the local system to the unpredictable future actions of the technical bureaucracy.

INCREASING FARMER INVOLVEMENT IN LARGE PUBLIC SYSTEMS

The main thrust of the above discussion regarding community irrigation systems is that assistance to these systems must be provided in a manner such that the basic responsibility for system investment and governance is not dislocated from the user community to a state agency. Now, I turn attention to the parallel need in the government sector to restructure the pattern of investment and form of governance of public systems, with the result that farmer involvement in critical irrigation development and activities is expanded.

The intensification of irrigated agriculture in public systems serves to draw more tightly together the traditional spheres of agency responsibility and community responsibility. In a loosely-operated (and extensive) system there may be a large, and amorphous, gap between where the agency stops being responsible for water investments and supervision and where the community assumes such. However, it appears that as water is used more intensively (i.e., becomes more scarce and valuable) both groups may act to fill the gap and, indeed, even to penetrate the traditional boundaries of the other. One sees this, for example, when farmer groups begin to patrol the delivery channels above their intake point to assure that adequate supplies will arrive. Or, when the agency begins to invest below the turnout in designing and constructing so-called tertiary facilities; structures previously said to be the responsibility of the users themselves. Such acts are themselves attempts to redefine the spheres of responsibility for each of the parties involved; though clearly, they work from opposing principles. One principle assumes that the gap is best filled by the agency extending its control and governance into the affairs of the tertiary unit; the other assuming that the solution lies in expansion of community governance to organize the unorganized gap.

While the matter still is unresolved, the evidence is accumulating that there is greater benefit in system performance to be gained from shifting the balance of investment and governance to increase the community role, as compared to extending the role of the technical agency. The indeterminant outcome of each of these approaches results, on the one hand, from the difficulty that bureaucracies encounter in obtaining and processing large amounts of location-specific information and sustaining the investments made and, on the other hand, the difficulties that can arise when large class differences exist in the local community.

The shift in investment and governance responsibilities is being explored in some nontraditional manners in several Asian countries today. Rather than simply looking for ways to involve farmers in more operation and maintenance (O&M) activities--that is, in a sense, to co-opt them for performing routine functions that the bureaucracy is unable to achieve--these innovative approaches are attempting to engage farmers in irrigation development tasks from which they were previously excluded and considered inappropriate contributors. In particular, attempts are being made to involve them not only in operating the irrigation works but also in designing the layout of the works at the tertiary level. The theory of this approach rests on two items: (1) the layout design at the tertiary level requires far more location-specific data than an external design team can hope to obtain, and therefore must rely on the knowledge and experience of local people with the physical and socioeconomic terrain they utilize; and (2) incorporating this indigenous knowledge results not only in a superior design, technically, but also a locally accepted one.

As with developing new strategies to assist community irrigation systems, strategies to achieve greater farmer involvement in irrigation development will require significant modifications in agency procedures and behaviors, if they are to be successful.[17] The ordering of design and construction steps, budgeting procedures, and measuring project performance are some of the fundamental agency processes that will require reexamination.

One of the most fundamental changes will be the timing of, and the effort to create new, or assist existing, local irrigation organization. The usual timing must be reversed--for example, rather than building new tertiary structures and then attempting to organize farmers to use the novel apparatus that is in place--the order should be to first organize farmers and then engage them in the processes of design layout and construction prior to their using the improvements.[18] Obviously, this new positioning of the organizational effort will have important implications for budgets, project timetables, and other standard procedures.

But it is not just a matter of sequencing; a much more serious staff commitment must be made to organizational tasks. Serious consideration must be given to utilizing staff with special training and experience in working with farmers to help evolve new or strengthen old irrigation groups. In most cases, simply adding this critical task to already busy irrigation engineers or agricultural extension workers will fail. Not only will the demands on their time limit serious attention to these matters, their lack of experience and skills will hinder success. While this may be more readily apparent for the engineering staff, it should also be noted that it often applies equally to extension staff. The basic task in working with irrigation groups is not disseminating information (the task for which extension agents are trained), but assisting in identifying group tasks, resolving conflicts and competition, imparting leadership skills, and fashioning appropriate organizational structure and procedures. These are difficult tasks and should not be uncritically assigned to whichever staff members happen to be "closest" to the farmers.

Identifying, recruiting, training, and utilizing such institutional staff as part of the irrigation agency is a complex task but one which a few Asian irrigation agencies are now exploring.[19] A major challenge of the 80's will be to examine and assess these experiences and, if they are judged productive, identify ways in which this approach could be used elsewhere. The approaches are being tried in relatively small national settings (the Philippines and Sri Lanka), and have not yet been considered in the enormous command areas of the subcontinent.[20]

Lastly, there is the need to restructure investment so that local communities are induced to add their own resources to those provided by the government. For example, rather than simply building tertiary structures in large public systems there is need to explore means for subsidizing or providing low-cost loans for local farmer groups to do this themselves.[21] The objectives of this approach would be two-fold; to enhance the total investment being made by matching government and community resources and, by inducing local investment, also enlisting local responsibility. But, such an approach is possible only if the technical agency is willing to allow farmer initiative in developing the facilities and the placing of "ownership" and governance of the facilities in local groups.[22]

CONCLUSION

In sum, while the major trend has been, and largely continues to be, toward state dominance in irrigation investment and management control, one need not assume that this is either inevitable or desirable. In an earlier period, community investment and management of irrigation works was the usual pattern, and even today persists as a major part of the irrigation sector in much of Asia.[23] Furthermore, within the government sector itself, there has long been an assumption that local communities had important irrigation functions to perform; recently, we are seeing an expansion of this view to

162

incorporate local groups in matters of system operation at the local level, and also in the decisions regarding the modification and elaboration of the physical structures and their layouts.

As with other aspects of agricultural development, irrigation development requires an appropriate fit between technology and social organization. In Asia, the transfer of large-scale irrigation technology has fostered the development of large and powerful technical agencies which often have acted to reduce local-level rights and responsibilities for irrigation management. Irrigation development policies have been insensitive to the rich tradition of community governance of, and investment in irrigation works that characterizes the Asian scene. The fundamental importance and magnitude of irrigation development required to increase food production and generate increased employment opportunities in Asia requires a reassessment of current approaches.

REFERENCES

[1]The classic, and controversial, statement on state power and irrigation development is Karl Wittfogel, Oriental Despotism (New Haven: Yale University Press, 1957). Excellent discussions of this thesis are found in E.R. Leach, "Hydraulic Society in Ceylon," Past and Present 15 (1959):2-26 and William P. Mitchell, "The Hydraulic Hypothesis: A Reappraisal," Current Anthropology 14 (1973):532-534

[2]One excellent description of these local arrangements is provided by Clifford Geertz, Negara: The Theater State in Nineteenth-Century Bali (Princeton: Princeton University Press, 1980).

[3]David Ludden provides important detail on these points for the case of Tamil Nadu, "Patronage and Irrigation in Tamil Nadu: A Long-Term View," The Indian Economic and Social History Review 16 (1979):347-365.

[4]For the case of Java see Anne Booth, "Irrigation in Indonesia, Parts I and II," Bulletin of Indonesian Economic Studies 13 (1977):45-77.

[5]See, for example, Donald W. Attwood, "Capital and the Transformation of the Agrarian Class System: Sugar Production in India" (Paper presented at the Conference on South Asian Political Economy, New Delhi, 1980).

[6]Such a system is described by Richard K. Beardsley, John W. Hall, and Robert E. Ward, "Japanese Irrigation Cooperatives," Irrigation and Agricultural Development in Asia, ed.

E. Walter Coward, Jr. (Ithaca: Cornell University Press, 1980), pp. 127-152.

[7]William W. Kelley, "Irrigation Management in Japan: A Critical Review of the Japanese Social Science Literature" (A report prepared for the Ford Foundation, 1982).

[8]Masakatsu Akino, "Land Infrastructure Improvement in Agricultural Development: The Japanese Case, 1900-1954, "Economic Development and Cultural Change 28 (1979):97-117.

[9]This clearly is the strategy of Indonesia's small-scale irrigation program as described by Oesman Djojoadinato, "Indonesia's Simple (Sederhana) Irrigation and Reclamation Program," Irrigation Policy and Management in Southeast Asia (Los Banos, Philippines: International Rice Research Institute, 1978): 25-30. In the Philippines new approaches are being implemented, see for example, Benjamin Bagadion, "Promoting Participatory Management on Small Irrigation Schemes: An Experiment from the Philippines" (London: Overseas Development Institute, ODI Network Paper, 1981) while the old strategy of converting community systems into agency systems sometimes persists in regions such as Ilocos Norte.

[10]While many individual studies could be cited to support this statement, a comprehensive view of one national setting is provided by Romana P. de los Reyes, Managing Communal Gravity Systems (Quezon City, Philippines: Institute of Philippine Culture, Ateneo de Manila, 1980).

[11]The limitation of bureaucracies in implementing irrigation development and managing irrigation systems are discussed by Robert Wade, "The System of Administrative and Political Corruption: Canal Irrigation in South India," The Journal of Development Studies 18 (1982):287-328 and M.P. Moore, "Approaches to Improving Water Management on Large-Scale Irrigation Schemes in Sri Lanka," Occasional Publication no. 20 (Colombo, Sri Lanka: Agrarian Research and Training Institute.

[12]Akira Tamaki, "The Development Theory of Irrigation Agriculture" Special paper no. 7 (Tokyo: Institute of Developing Economies, 1977).

[13]Two case studies of this program of subsidies to community irrigation are found in Anwar Hafid and Y. Hayami, "Mobilizing Local Resources for Irrigation Development: The Subsidi Desa Case of Indonesia," Irrigation Policy and Management in Southeast Asia (Los Banos, Philippines: International Rice Research Institute, 1978), pp. 123-134.

[14]A good description of a Balinese subak is provided by Clifford Geertz, in "Tihingan: A Balinese Village," in Irrigation and Agricultural Development in Asia, ed. E. Walter Coward, Jr., (Ithaca, NY: Cornell University Press, 1980), pp. 70-90. The zanjeras of the Northern Philippines have been discussed in Henry Lewis, Ilocano Farmers (Honolulu: University of Hawaii Press, 1971), E. Walter Coward, Jr., "Principles of Social Organization in an Indigenous Irrigation System." Human Organization 38 (1979):28-36; Robert J. Siy, Jr., "Rural Organizations for Community Resource Management: Indigenous Irrigation Systems in the Northern Philippines" (Ph.D. dissertation, Cornell University, 1981). The muang-fai systems of Northern Thailand are detailed in Abha na Ayutthaya, "A Comparative Study of Traditional Irrigation Systems in Two Communities of Northern Thailand" (Bangkok: Chulalongkorn University Social Research Institute, 1979).

[15]Perhaps the most apparent historical case of this type is that of the community tank systems in South India. On the current state of these systems, see the Proceedings from the Workshop on Modernization of Tank Irrigation held at Perarignar Anna University of Technology, Madras, February 10-12, 1982.

[16]Some information on irrigation in North Sumatra is provided in Richard P. Lando, "The Gift of Land: Irrigation and Social Structure in a Toba Batak Village" (Ph.D. dissertation, University of California, 1979).

[17]Needed modifications in agency procedures are discussed by Felipe B. Alfonso, "Assisting Farmer Controlled Development of Communal Irrigation Systems," in Bureaucracy and the Poor: Closing the Gap, eds. David C. Korten and Felipe B. Alfonso (Singapore: McGraw Hill International, 1981).

[18]These matters are discussed with regard to Philippine experiences in Benjamin U. Bagadion and Frances K. Korten, "Government Assistance to Communal Irrigation in the Philippines: Facts, History and Current Issues," Philippine Agricultural Engineering Journal 10(1979):5-9.

[19]Currently, such innovative approaches are being tried in Indonesia, the Philippines, and Sri Lanka.

[20]References to the Philippine experiences are cited above. The Sri Lanka program is presented by Laksham Wickramasinghe and Edward J. Van der Velde, "Action Research into Farmer Participation in Irrigation System Management: A Sri Lankan Experiment" (Paper presented at International Conference on Field Research Methodologies for Improved Irrigation Systems Management, Coimbatore, Tamil Nadu, India, September 15-18, 1981). Robert Wade has raised the matter of the relevance of

the Philippine approach for India, "Collective Responsibility in Filipino Irrigation: A New Approach," Economic and Political Weekly 16 (1981):A-99 to A-102.

[21]The Japanese case discussed by Masakatsu Akino (see above) is instructive on this point.

[22]As analysts have noted, farmer interests in such terti-ary-level investments is unlikely if the irrigation agency is unable to effectively operate the main system. See R. Wade and R. Chambers, "Managing the Main System: Canal Irrigation's Blind Spot," Economic and Political Weekly 15: (1980); and Anthony Bottrall, "The Management and Operation of Irrigation Schemes in Less Developed Countries," Water Supply and Management 2(1978):309-332.

[23]One thinks of both the ancient subaks of Bali and the modern tubewell systems of India and Bangladesh.

Subject Index

planning, 51

Repeasantization, 131, 132, 133
Research, 20
  centers, *See* Agricultural research
Resource deterioration, 18
Rhode Island, 47
Risk, 8, 86, 88, 96, 101, 102, 117, 120
  attitudes toward, 94, 101
  aversion, 85, 86, 87, 97
  evaluation, 90, 96
  perceived, 117
  preference, 88
Russia, 44

Sahara, 17
Sahel, 18, 24
Saudia Arabia, 129
Scale-neutral, 9
Semiproletanianization, 123, 131, 133
Senegal, 23, 24
Shadow prices, 54
Slash-and-burn, 18
Social
  and political barriers, 9
  norms, 10
Socioeconomic constraints, 130
South America, 32
South Korea, 129
Soviet Union, 18
Specialists, 62
Subaks, 159
Sumatra, 159
Sweden, 47

Tailoring technology 4, 25
Taiwan, 17, 25, 77, 129
Tanzania, 60, 117
Technological
  change, 27, 40, 136
  fit, 147, 148
  inferiority, 127
  interventions, 136
  mastery, 7, 11
Technologies, improved, 19, 58

Technology, 2, 20, 23
  appropriate, 4, 38, 125, 150
  development, 39
  impacts, 8
  package, 30, 36, 40
  tailored, 4, 25
  transfer, 77
  transfer constraints, 24
  transfer of, 2, 3, 10, 11, 26, 75
  type, 20
Thailand, 8, 17, 34, 44, 88, 92, 94, 96, 97, 98, 130, 159
Third World, 23, 120, 125, 128, 131, 132, 134, 135, 136
Tilapia, 113
Tractors, 138
Training, 26, 27, 80, 81, 82
Tunisia, 8, 88, 92, 94, 96, 98, 102

UNDP, United Nations Development Programme, 46, 47
UNIDO, United Nations International Development Organization, 53
United Arab Republic, 77
United States, 20, 22, 26, 43, 36, 72, 77
  Agency for International Development, *See* AID
Universities, U.S., 22, 26
Urban Sector, 34
USAID, *See* AID

Varieties, 19, 24
Venezuela, 17, 129
Vietnam, 156
Vision, 30, 31, 35, 39

Women, 149
Working class, 127
World
  Bank, 55
  food demand, 18
  system, 9

Author Index

172

About the Contributors

GEORGE H. AXINN is Professor of Agriculture Economics and Assistant Dean of International Studies and Programs at Michigan State University, where he also serves as Adjunct Professor of Education, Associate Coordinator of the Center for Advanced Study of International Development, and Chair of the Farming Systems Research Group.

NYLE C. BRADY is Senior Assistant Administrator for Science and Technology in the Agency for International Development. From 1973-81 he was Director-General of the International Rice Research Institute in the Philippines. Prior to that, he served on the faculty and was Director of the Cornell University Agricultural Experiment Station. Dr. Brady is the author of a landmark textbook in the field and has published numerous research reports and journal articles.

FREDERICK H. BUTTEL is Associate Professor of Rural Sociology at Cornell University. Having received a Ph.D. from the University of Wisconsin, he has written widely on environmental and agricultural issues.

E. WALTER COWARD, JR. is Associate Professor of Rural Sociology and Asian Studies at Cornell University. He received his Ph.D. at Iowa State University. For a number of years he has been actively engaged in research on issues of irrigation development in Asia.

WILLIAM L. FLINN is Executive Director, Midwest Universities Consortium for International Activities and Professor of Rural Sociology and Sociology at The Ohio State University. He has conducted research and consulted in a number of developing countries.

TERRY ROE is Professor of Agricultural and Applied Economics at the University of Minnesota. His research in foreign economic development focuses on development planning, food policy and technological change. He has long-term experience in Tunisia and had rendered short-term experience in numerous countries.

W. W. SHANER is Associate Professor in Industrial Engineering at Colorado State University. He holds a B.S. in Civil Engineering from Iowa State University, an M.B.A. from Harvard University, and a Ph.D. in Engineering-Economic Planning from Stanford University. He has worked extensively with the governments of Ethiopia and Peru and has contributed to development projects in a number of other nations.

JAMES R. SIMPSON is Professor and Livestock Marketing Economist at the University of Florida. He has long-term foreign experience in Paraguay, Chile, and Costa Rica, and has rendered short-term service in numerous other countries in Africa, Asia, Oceania and Latin America.

ROBERT D. STEVENS is Professor of Agricultural Economics at Michigan State University. Receiving a Ph.D. from Cornell University, his international experience includes service in Southeast Asia, Pakistan, and Bangladesh.

JAMES A. STORER is Director, Office of Fisheries Affairs, Department of State, Washington, D.C. Dr. Storer received a Ph.D. in Economics from Harvard University. He has served on the faculty and as a Dean at Bowdoin College. Twice a Fullbright Scholar to the University of the Philippines, he has held administrative positions with the National Oceanic and Atmospheric Administration, as well as the Food and Agricultural Organization.

DONALD R. STREET is Associate Professor of Economics at Auburn University and a member of the Advisory Committee of the International Center for Aquaculture. He received the Ph.D. degree in Resource Economics from The Pennsylvania State University. He has completed fisheries economic studies in central and South America, the Caribbean area and Africa. He teaches courses in Economic Theory and Economic Development.

GREGORY M. SULLIVAN is Assistant Professor of Agricultural Economics at Auburn University. He received the Ph.D. from Texas A&M University and has served overseas with the Peace Corps in Ghana, as a staff member on a cattle marketing project in Tanzania, and as a marketing consultant in an Indonesian aquaculture development project.